THE GREEN PLANET

THE GREEN PLANET

The Secret Life of Plants

Simon Barnes

WITNESS
BOOKS

CONTENTS

INTRODUCTION

What have plants ever done for us?

That's a question that seems hard to answer as you walk, or more likely drive, through the streets of just about any city in the world. It seems as if we have cut our ties with plants altogether: that we no longer need them to survive, as our ancestors did. Certainly we seem to be running the planet on that assumption.

In general terms we accept that plants are part of life and therefore probably a good idea. But this doesn't go nearly far enough. Plants *are* life. Without plants life on planet Earth is not possible and never would have been possible.

Every morsel of food that you take into your system depends on plants. Every plate is ultimately a plateful of plants, not just the salad garnish. Our domestic animals convert plants into meat. (Beef cattle consume between six and 25 times as much food in a lifetime as they actually produce.) The food chain in the ocean ultimately depends on phytoplankton, minute oceanic plants. Even fungi are consumers of plants.

But food is only the most obvious part of our dependence on plants. They also supply the oxygen we breathe. We take in oxygen and breathe out carbon dioxide; plants take in carbon dioxide and give out oxygen, so we can – but alas we mostly don't – look on the rainforests as a great oxygen factory.

Plants also play a central part in the cycle of water, taking water from the ground and firing it at the sky, where it forms clouds and falls as rain. So plants provide food, air and water, all of which are fairly useful when it comes to human existence.

We like to think that twenty-first-century human beings have cut their ties with nature, creating a world that no longer depends on such retro things as trees. But the modern world is still mostly powered by plants. The Industrial Revolution, the greatest change humans have made to the planet since the invention of agriculture 12,000 years earlier, was powered by coal, and coal is just another form of plant, mostly fossilised trees. Our dominant source of power is now oil, and oil is also fossilised life: plants, algae and other marine organisms. Now we are trying to change our dependence on oil by developing biofuels: fuel made from living plants.

OPPOSITE
A working environment: an American painted lady butterfly takes nectar from a coneflower while providing a pollination service to the plant, New Mexico, USA.

The way we have used this plant-based power has created the biggest crisis in human history: climate change. Global temperatures continue to rise inexorably, routinely triumphant over political platitudes. But plants help to control that, removing carbon dioxide from the air and storing carbon: acting as the planet's air-conditioning system. There are two ways of getting climate change under control, and they need to be done hand in hand: stop using fossil fuels and treasure plants as never before.

Plants, then, are central to every moment of our lives – but that's only half the reason for celebrating them in this book. In traditional nature documentaries and in most books about wildlife, plants play a supporting role, creating a picturesque background, providing the food for the cast and allowing them to breathe.

We are animals ourselves; it is natural to turn to our fellow animals when we seek to understand the world we live in, and naturally we love action and drama. We are enthralled by hunting lions, mating elephants, dancing birds of paradise, the social lives of wolves and the loneliness of polar bears. But advances in both science and photography have opened up the kingdom of plants as never before, revealing other kinds of drama, struggle, enmity, co-operation, invention and even hunting skills.

Many plants live on a timescale that is incomprehensible to us, having an existence of a few short weeks or sometimes thousands of years. The complex mysteries of the way that plants live have been revealed in moving pictures in the televised version of *Green Planet*, and in still images here in these pages.

In both these works plants are put into the foreground, while the animals they share the planet with take the supporting role. But as the complexity of life is revealed by more and better science, more and better ways of making images, we can see with uncompromising clarity that the lives of plants and animals are inextricably entangled, enabled and enriched.

We take this journey through the lives of plants in easy stages, starting with the incomprehensible luxuriance of life in the tropics, and then moving sharply on into the deserts, the places in which it would seem that life of any kind is impossible. From there we look at the astonishing contrasts of the seasonal lands, the way a landscape changes almost unrecognisably from one end of the year to the other, and then we move into the water: where all life began, to which a good deal of life has returned.

OPPOSITE
Making much of little: Mexican giant cardon cactus and a boojum tree, Baja California, Mexico.

OVERLEAF
Glorious excess: tropic rainforest, Costa Rica.

Humans began altering the planet for the convenience of their chosen plants with the dawn of agriculture and since then the planet has changed as completely as it did when the meteor struck the Earth 65 million years ago and did for the dinosaurs. So in the final chapter we look at plants in the context of humanity.

There are plenty of harsh truths to be faced here, as we all know. But in the midst of it all there are people trying to set things right. We find initiative, resourcefulness, optimism and joy. We are at the crossroads and humanity has a choice of routes. When we look closely at plants and how they work, we can see the correct choice without any problem whatsoever. All we need to do is make it.

TROPICAL
WORLDS

It begins with light. This day began with light, as all other days do. Once there is light, life – everything – can begin. Light is everything. Light is life, because light is food. We feast daily on light: and so does every pig and gorilla and cockroach that walks on the surface of the Earth, and so does every crab and tuna and blue whale that lives in the ocean.

It starts with light because of plants. The most ferocious species in the animal kingdom depend, in the end, on plants and, since plants depend on light, light is what drives the lives of almost everything that lives on the planet. The extraordinary rule-proving exception is the life associated with the hydrothermal vents that lie deep in the ocean, and here a complex range of life subsists ultimately on the energy that comes from the Earth itself. But this is a tiny and closed society. For all practical purposes it is the light – the light that comes from the sun – that makes all life on Earth possible.

It is so humdrum, such an everyday sort of miracle that we forget its miraculous nature. But the fact is that plants turn light into food. It's as if you could sit in the sun with your hand extended, and after a bit you find yourself holding a cheese and tomato sandwich. What plants do is no less fantastic than that. They make food for themselves. Food, for plants as for us animals, is fuel: and the plants use the food they make to power themselves, sometimes to dizzying heights, sometimes to survive for thousands of years, sometimes to become scraps of life so small and ephemeral we scarcely notice them. They need the fuel to make more of themselves, to reproduce themselves, sometimes in bizarre fashion, sometimes employing the help of members of the animal kingdom.

The process of turning light into food is called photosynthesis and it is the one natural process that makes all others possible. A plant takes in carbon dioxide through its leaves and water and minerals through the hairs on its roots. The sunlight powers the chemical process that creates glucose, the basic fuel in the life of plants. There is also a waste product that is excreted into the air. This waste product is oxygen – so plants don't just give us food, they also give us the air that we breathe.

The most determined human carnivore that ever lived depends as much on plants as the most resolute vegan. An obligate carnivore species like a lion depends on plants as much as the wildebeest who eats the grasses of the savannah. No grass means no wildebeest means no lion-food means no lions. The fungi that live all over and most often under the earth also depend on

OPPOSITE
The light and life of the rainforest is almost all in the canopy.
Only five per cent of the light reaches the forest floor here in
Malaysian Borneo.

plants. Unlike plants, they can't make their own food but, like us animals, they get their energy, directly or indirectly, from plants. Plants eat light. We humans are creatures of the daylight, poorly equipped for dealing with darkness and night. We have always equated light with life and goodness, darkness with death and evil.

We can be a little thrown, then, should we ever have the privilege of walking in rainforest.

The compound word rainforest was a specialist term until 1984. This was the year in which *The Living Planet* was broadcast. It was the middle part of Sir David Attenborough's *Life* trilogy, which began with *Life on Earth* and ended with *The Trials of Life*. The theme of *The Living Planet* was ecology, also not a word in common use back then. It was in the fourth episode that viewers encountered the great forests that encircle the middle part of the world. Sir David wrote in his book of the series: 'Nowhere is there more light, warmth and moisture than in West Africa, Southeast Asia and the islands of the Western Pacific, and South America from Panama across the Amazon

Basin to southern Brazil. As a consequence these lands are blanketed by the densest and richest proliferation of plants to be found anywhere in the world. Technically it is described as evergreen tropical rainforest. It is more widely known as the jungle.'

Not any more, it isn't. That episode, above all others, caught the public imagination. A rough and ready understanding of the richness of its life, and the inexorable horrors of its continuing destruction are part of twenty-first-century life. The idea of saving the rainforests has gone round the world. When you walk in rainforest for the first time you have certain expectations. You assume you will step in among the trees and be

ABOVE
Jungle lover: David Attenborough visited the Ecuador rainforest during the making of his 1984 series The Living Planet.

overwhelmed with life as never before; that you will have an instantaneous experience of biodiversity that will fulfil a lifetime's ambition and impart a lifetime's worth of understanding. But that's not what it's like.

If you walk across the floor of a rainforest, you find yourself walking in darkness, or near-darkness, even at noon. You will find plenty of insects eager to keep you company – people who work for extended periods in rainforest often wear a bee-veil to protect their faces – but the great glorious vistas of complex interdependent life forms will mostly be out of your reach. It's all a bit above your head, in the most literal way. You can hear sudden piercing calls from birds you can't see, the occasional crash of some creature moving through the trees far above you, but mostly you will feel lost, perhaps even a little depressed. Occasionally, you will encounter the tiniest ray of sunlight squeezing in through a gap, so sudden and so bright that it looks like a biblical miracle, a pillar of light. And you look up: up at the canopy of the forest through which the light had passed, and you feel like Alice in *Alice's Adventures in Wonderland*, forever locked out of the beautiful garden. If only, you think, I could fly, or I could climb with the weightless ease of monkeys. If only I could leave the forest floor and be up among the treetops – for up there, there is light.

ABOVE
A wonder that never gets stale: David Attenborough filming in
rainforest at La Selva Biological Station, Costa Rica.

In the first days of Narnia, as told in CS Lewis's *The Chronicles of Narnia*, the land is so rich and fertile that anything that anyone plants, on purpose or by accident, will grow: a fallen iron bar grows into a lamp-post; dropped coins become gold trees and silver trees; and, overnight, a planted toffee grows into a toffee-tree with sweet fruit and papery leaves. It feels just like that in tropical rainforest: a place so quick with life that you feel your boots might sprout into leaf while you sleep, and that if you stand still for too long you will be covered in epiphytic plants. Take away the trees, and you'd fancy that anything in the world might grow here instead: but that's not what happens.

There is a dismaying circularity about tropical rainforest. It's fertile because it's fertile, it's wet because it's wet, it's full of life because it's full of life. Take away the trees – and alas that has been done again and again all

ABOVE
Impossible fecundity: a stream crosses the forest floor at the
Lambir Hills National Park in Malaysian Borneo.

over the world – and you find pretty poor soil. The luxuriance of the forest doesn't come from the hidden richness of the soil beneath, because there isn't any. It comes from itself: the luxuriance of the forest is what makes the luxuriance of the forest possible: a luxuriance established and maintained across millions of years.

The warmth and moisture make a very promising start to the processes of life. The reason that the tropical forests have so much life is because they have stood undisturbed for so long, unchanging, allowing life to create more life. And the numbers still have the power to boggle and beguile: tropical rainforests cover two per cent of the land surface of the planet and support more than half of the world's species of animals, plants and fungi. Some estimates put the figure still higher: maybe as much as three-quarters. When

ABOVE
Reaching for light: a view from a canopy platform in
Lambir Hills National Park.

a habitat is both undisturbed and unfathomably rich, living things have the opportunity to explore apparently fantastic ecological niches and evolve to exploit them with bewildering efficiency. The plants, fungi and animals frequently take what to us seem bizarre shapes, forming extraordinary relationships of interdependency. It is certain that there are millions more species yet to be discovered here, living life in ways that are not only more fantastic than we imagine but more fantastic than we are capable of imagining.

In less extreme environments, growing plants deposit humus and make the soil richer. That doesn't happen in tropical rainforest because the circumstances that make for rapid growth – heat and moisture – make for equally rapid decay. The incessant rain washes nutrients away. There is no point in a tree developing long roots to reach hidden riches of the soil, because there are no hidden riches. Instead, the roots are shallow, and take minerals from the forest floor: that which comes from decomposing leaves and animals. Thus the forest fertilises itself: take away the trees and the process stops.

A plant must cycle water through itself to keep alive: as blood flows through our veins, so water flows through the vascular system of plants. The leaves not only deal with the essential task of photosynthesis, they also let out water in the form of vapour. They pick up water through the root hairs, it flows upwards through the plant – sometimes reaching colossal height, 80 metres and more in the trees that stand proud of the canopy – and pass the water out into the atmosphere. When you add together all the leaves on all the trees in the rainforest, you have a great deal of water vapour. This forms clouds which fall on the forest in the form of rain. Rainforest not only fertilises itself, it also waters itself.

So a rainforest is, to a large extent, a closed system: its energies turned in on itself. The stability of the system is at the heart of its extraordinary riches: rainforest untouched by human hand have been essentially the same for 55 million years: only ten million years later than the last *Tyrannosaurus rex*. And that's how it feels to the entranced visitor, craning upwards in the darkness of the forest floor: forests that make human civilisations look puny and human lives seem petty. It doesn't seem, in the stillness of a forest morning, as if anything could ever change here.

OPPOSITE
Life's complexity: a giant liana reaches for the canopy in the jungles of Panama.

OVERLEAF
You never forget that jungles are about moisture: dawn mists over the forests of Danum Valley Conservation Area, Malaysian Borneo.

But the fact is that the forest is changing all the time: if it didn't it wouldn't be forest, it would be dead. The stability of the forest is based on incessant motion. A bicycle is only stable when it is moving: the forest lives because it is always moving. The reason it is able to stay the same is because it is always changing. That system of perpetual change is all to do with the light.

So let us go back into the darkness, down on the forest floor. Under the closed canopy of the forest only about two per cent of the sunlight reaches the ground. Any plant that starts to grow from seed down here is quite literally starved of light: no light, no food, no life. Seeds that do germinate grow with terrible, almost hopeless slowness: a stunted sapling may have taken ten years to become a midget, a pace of life more often associated with deserts than with rainforest. Had it been granted the gift of light, it might by now have been halfway to joining the company of giants. What's more – what's worse – is that by standing up from the ground they make themselves vulnerable to the browsing mammals that walk the forest floor. It seems hopeless. The most life-thronged environment on the planet is incapable of supporting life at ground-level. It seems like a monopoly for the trees that have grown tall and taken all the light: nothing else, nothing at all, is allowed in. You're either up in the canopy or you're dead.

The forest floor is full of seeds that can't germinate and seedlings that seem stuck in infantilism. It all stays like that, apparently for ever: and for decades the terrible stability remains, the seedlings wilting and dying, the seeds losing their vitality, more seeds falling, some germinating with apparent hopelessness.

And then there is light. Light in the form of lightning. The rain from the clouds created by the action of the trees beneath will fall sometimes as gentle showers, at others times in fits of primal ferocity. The most ferocious electrical storms are part of the routine of forest life, and when lightning strikes it has to come down somewhere: the shortest way down for preference. As lightning seeks out church steeples or the top of the Empire State Building, so in the forest it makes its strike on the tallest trees. That will often mean the emergents: the trees that stand tall of the canopy. The kings of the forest are the most likely to be toppled, and the higher they stand, the more disruption they cause when they hit the ground. It's been reckoned that 40 per cent of the tall rainforest trees are felled by lightning: dying from storms that they have helped to create, the same sort of storms that had watered them throughout their long lives.

OPPOSITE
TOP LEFT *The race begins: new sprouted seedlings start to grow towards the light in Malaysian Borneo.*

TOP RIGHT *Cassowary's work: this blue quandong seedling is one of 70 tree species that depend entirely on the southern cassowary to spread their seeds.*

MIDDLE *Take the fruit but spread the seeds: a collared aracari at work in Central America.*

BOTTOM *A dung-pile of life: the cassowary of Papua New Guinea and Australia eats fruit and disperses the seeds. This two-kilogram pile contains more than 300 seeds.*

Down they come, creating a giant wound – and with it, a host of opportunities. In the new gap everything dries out: the sun is ferocious and the protection gone. It's as if all the windows of the greenhouse had been broken, the still, moist atmosphere dissipated. As a result of this new dryness, fires will break out, as destructive as they are cleansing.

There are no disasters in nature. Destruction is another word for opportunity. As the emergent giant falls and the fires blaze around the clearing, it's as if the forest had thrown a double-six and started again. A sizeable area of the once-dark forest floor has been flooded with light: fierce, intense tropical light, interrupted only by warm nights and refreshing douches of rain. After the lightning comes the life.

We humans respond to light: light makes us cheerful and more positive, more filled with a sense of wellbeing. We can take that feeling and multiply it a thousand times if we want to understand the way that plants respond. If we

OPPOSITE
TOP *Cassowaries take fruit of all kinds; here's a small selection from the forests of North Queensland, Australia.*

BOTTOM *Endless forms: fruit and seeds from Royal Manas National Park in Southern Bhutan.*

ABOVE
Lightning kills trees, and each death is an opportunity for thousands: a storm in Dzanga-Ndoki National Park in Central African Republic.

can go back to the *Chronicles of Narnia*, it's as if the creatures who had been turned to stone by the evil of the White Witch had been brought back to life by the breath of the great lion Aslan. Light, the gift of life, has been granted by the lightning, and now comes the overwhelming response. The slow-motion seedlings turn from tortoises to hares and start to gallop towards the sun. Seeds that have been lying dormant across decades turn from old seeds into new plants.

There is opportunity for all: but there is success only for comparatively few. Which species, which individual plants are best capable of exploiting this opportunity? What is the best strategy for victory? Is there one strategy or many? Victory is about becoming an ancestor: growing enough to reach maturity and

ABOVE
To the winner, the ultimate prize: unlimited access to light.
A tualang tree emerges from the canopy in Malaysian Borneo.

reproduce, with your genes undying in your progeny. It's fiercely competitive and the prize is immortality, at least immortality in the form of parenthood. Most of your neighbours are your enemies. In short, it's a jungle out there.

For some there is the sprinter's option. The choice of a short life and a merry one is perfectly exemplified by a plant known as hotlips, or, still less tastefully, hooker's lips, not a reference to the great nineteenth-century botanist Joseph Hooker. The name, presumably referencing the 1970 film *M*A*S*H* and the character Major Margaret 'Hot Lips' Houlihan, played by Sally Kellerman, is irresistible: the plant seems to be an almost deliberate burlesque of what plants should be, what Uncle Monty, in the film *Withnail and I*, called 'prostitutes for the bees'.

OVERLEAF
A study in diversity: Carro Colorado Island, Central Panama,
has more than 500 species of trees.

The hotlips plant grows in the tropical rainforests of Central and South America, and it does so fast, stealing a march on those with long-term ambitions. It germinates in one of those forest gaps made possible by the fall of a giant and leaps into flower as early as possible, trumpeting out its existence to the potential pollinators of the forest with almost hysterical urgency. Everything about the plant is about instant gratification: the window to operate in is narrow and closing fast, as the more robust plants all around begin their own campaign to reach the light. We think of plants as things that live at a speed somewhere between slow and stop: we need to adjust our animal-chauvinist understanding of the way time works. As hotlips gets on with the job of growing and flowering, we are watching a sprinter in action: the Usain Bolt of the forest floor. Lightning began the process, making the gap and the opportunity: this is a plant moving at the vegetative equivalent of lightning speed.

ABOVE
A target that's hard to miss: a brown violetear hummingbird homes in on the plant known as hotlips in Costa Rica.

OPPOSITE
A study in scarlet: the hotlips plant, named for its bight bracts, is an attention-seeker.

The luscious red lips are not petals; technically they are not a flower at all. They are bracts, modified leaves: we see the same thing in the orange bracts of the poinsettia, an indoor plant much grown at Christmas time. But, in both species these bracts serve the function of flowers: as a come-hither to pollinators, one that is impossible to miss. Inside the bracts of the hotlips plant – between those lips – there are small creamy white flowers, modest things when compared to the bravura advertisement that brings the pollinators in. Here they are available for hummingbirds and butterflies, to take the offering of nectar and then transport the pollen to the next flower. Flowers allow plants of the same species to exchange genetic information – to breed – and many plants employ an animal intermediary to do the job.

Once hotlips has been satisfied, the fruit will form: small blue-black berries. These are dispersed, mostly by birds, and the seeds will land in other parts of the forest, to remain underground for year after year, sometimes to all lose all potency. But the odd one or two will win the forest lottery and germinate when the forest next receives a wound, when the next giant falls

and the light floods the newly made clearing. Then a new generation of hotlips will respond as eagerly as did the parent plant. Their ambition is not great, in forest terms: if they make a shrub two or three metres high, they are exceptional, and they live only for a handful of years. But, looking at it another way, their ambition is as great as anything that ever lived: which is to become immortal. That is to say, to become an ancestor, to pass on those immortal genes and live forever by means of offspring.

The strategy of notice-me flowers (and bracts) is taken up on the far side of the world by another plant, one that, like the hotlips plant, has become a favourite among green-fingered greenhouse cultivators. The bat flower of Southeast Asia also crops up in the understorey: able to survive in low light when there is an incomplete canopy. The flowers are purple, near black, sometimes as much 30 centimetres across and with 'whiskers' up to 70 centimetres long. To see similarity to a bat is a little fanciful, but there are eager gardeners who love to produce them for Halloween. In the wild, they are making a strong statement: and that statement is 'look at me' as they seek to bring pollinators in, pollinators who might otherwise be distracted by the riches that grow high up in the canopy.

Hotlips is an opportunist and the plant is all about excess: excessive speed and excessive advertisement. It works: the plant flowers and fruits and dies, after which its seeds lie dormant, waiting for the next opportunity. While hotlips is living its brief moment in the spotlight of the sun, other plants with more long-term ambitions are growing towards the light at a more sedate pace.

If plants like hotlips are sprinters, the great giants of the forest are marathon runners. That leaves a gap for the middle-distance runner: for the contender who has something of the speed of sprinter and something of the long wind of the endurance specialist. The champion of the half-and-halfers is the balsa tree.

We are most of us familiar with the idea of balsa wood as a substance. That familiarity gives us the clue to the success of the balsa tree. Balsa wood is used by enthusiasts for making models, especially model aeroplanes. It can be cut with ease, readily takes any shape the modeller chooses, and above all, it is phenomenally light; *balsa* is Spanish for raft or float. It is so light and supple that it challenges our sense of what is right when we hold it. It feels

OPPOSITE
Look at me: the bat flower of Southeast Asia stands out
from the background in order to attract pollinators.

like wood, and yet it doesn't: it's too light, too airy and much too easy to bend and snap. It feels almost like a synthetic product thing: too peculiar to be wholly natural.

This is wood that doesn't feel like wood – and that is the secret of the balsa tree's success. By using this unwoody kind of wood, the balsa makes a real tree – but it's still not entirely like a real tree. It is a tree that carries its own weight more than a little precariously: a tree not quite up to the standards and responsibilities of the forest giants. The balsa is kind of a half-baked tree: but that strategy of being half-baked is a kind of genius.

The other trees of tropical forest have hard wood that grows very slowly. Time is needed to create these immense and powerful structures. When they're cut down they reveal dense wood, with the annual growth rings – those that you count to learn the age of the tree – very narrow and close together. The narrowness is a truthful indicator of the slow rate of growth: girth that increases just a little every year. That way the tree is able to carry its own weight with ease and withstand many storms in the course of a lifetime that can be measured in centuries. Trees like this are committed to the strategy of slowness.

The balsa goes for speed. That's speed as a tree rather than as a cheetah understands the term, but it's about speed all the same. David Attenborough was filmed in front of a balsa tree ten metres high. It was a year old; at the same age, a hardwood tree would stand a few centimetres. The secret of the speed is all about water: all about the balsa tree's elevated talent for taking on water. Plants capable of moving water about inside themselves are called vascular plants, making an immediate call to be compared to our own animal system for shifting blood. Plants absorb water through the root hairs by the subtle process called osmosis. This is then moved right through the plant, but without any beating heart to do the job. How do they do it?

Here's a homely experiment that will give you the idea: leave a towel on the edge of the bath with the bottom inch in the water. Come back a few hours later and the whole damn towel is wet. We call it wicking: it happens to our trousers when we walk through wet grass, the liquid defying gravity and climbing relentlessly upwards towards your crotch. Plants employ the same sort of system: the water rises up through the inner bark or xylem by what is called capillary action. Water will climb unprompted up a tube, if the tube is

OPPOSITE
The balsa and its clientele.

TOP *A white-necked Jacobin hummingbird steals nectar from the base of the flower without supplying pollination services.*

MIDDLE LEFT *A kinkajou prepares to take a draught of nectar.*

MIDDLE RIGHT *A woolly opossum is a committed pollinator.*

BOTTOM *A greater spear-nosed bat – an omnivore and the second largest bat in the New World – homes in on the balsa flowers.*

narrow enough. It's all to do with intermolecular forces: and to those of us accustomed to the routines of gravity, it looks quite a lot like a miracle.

The balsa tree is very good at drawing up a great deal of water and secreting it about its tissues in cells: that's what gives balsa wood its holey nature. The tree grows this loosely made wood at express pace. There are times in a rainforest when human visitors fancy they can almost hear the trees growing: the balsa, well, you can almost see it growing.

It's not a great tree, but then it's not trying to be one. It's vulnerable to wind, it's a weakling compared to its much older neighbours. But it grows right up to the canopy, making a good 30 metres in a few years. It is in a rush to grow; it is also in a rush to reproduce, and it does so faster than any other tree in the forest.

Balsa trees can't supply the great riches of fruit that other trees manage when the time comes, but they compensate for their own inadequacy by making sure that they have very little competition. In the drier times of the year, most of the big trees take an energy-saving option. They slow down, they shed leaves to reduce water loss (the water vapour that is passed out from trees to form those rainforest clouds is transpired from the leaves). That is when the balsa makes its play. It sucks up water eagerly from the soil and employs its energies to make flowers. They are hugely attractive to the creatures of the forest, not because they are the most spectacular flowers the rainforest has ever seen, but because there isn't much competition. For many it's the only game in town.

The balsa trees put out big white flowers, with an inch-deep cup of nectar: the water-absorbing balsa can afford to be profligate with liquid. In the dry season each balsa tree is a little oasis. In six weeks a single tree can produce 2,500 flowers with a total of 15 gallons of nectar – enough to fill the tank of a family car. The flowers open for business in late afternoon and are much visited in the early part of the night. They need to cross-pollinate if they are to fruit – that is to say, to get pollen from another plant: only 15 per cent of self-pollinated flowers produce fruit, and these fruit are very poor in seeds. The balsa tree requires the services of active and committed animals that will accept the gift of nectar (and pollen as food) and as they pursue their own ends, they will incidentally serve the plant by performing the services of pollination. Pollen is a powder that contains male reproductive material. In

order for the plant to reproduce, the pollen needs to get from the stamens, the male reproductive organs, to the female reproductive organs, which in flowering plants are called pistils. We will come up against pollen again and again as we make this journey though the life of plants. Many animals perform sex in what seem to us complex, bizarre and even perverse ways; in the same sort of way, plants often send pollen to pistil in ways that startle the human imagination, and make us wonder how such extraordinary things could develop. The answer to that question is always the same: time. Not time as we understand it, in days and weeks and the span of a single human life time, but in the form of Deep Time: time measured with more noughts than our human brains can cope with. You would not look at a balsa tree in flower and say: well, obviously this tree reproduces itself with the help of a

ABOVE
Bounty of the canopy: a red-headed barbet feeds on cecropia
fruit in northwest Ecuador.

vegetarian carnivore with a prehensile tail and a ridiculously long tongue. But that's how it works.

What creatures visit these vast blooms and, refreshed by deep draughts of nectar, do the job of pollination with such efficiency? Scientists have traditionally assumed that it was bats. This was a sensible idea, but hard to prove: it's not easy to spend a night in the canopy of rainforest to check this out. But advances in technology have allowed humans to see what is going on from a distance, and researchers have found that, yes, bats do indeed visit the balsa flowers and drink, and fly off with a light dusting of pollen. You'd wonder if that was enough to do the job with the efficiency the tree needs to reproduce itself, however. And, it's been discovered by more recent research that the trees have another and much more eager pollinator.

This is the kinkajou: a small mammal that lives in the canopy of Central and South American rainforest. Rather unexpectedly, it is a carnivore, in that it is a member of the order Carnivora; but, like the panda, it is a largely

ABOVE
Plants growing on plants: a zebra bromeliad in the canopy,
Tiputini Biodiversity Station in Ecuador.

vegetarian carnivore – and this is not a contradiction so long as we are talking about taxonomy. Kinkajous come out at night, which is the best time to avoid predators like harpy eagles. When they get active on nights at the drier time of year, they find the balsa flowers have obligingly opened for them.

Kinkajous are not vegetarian on principle, and will take eggs and young birds when they find them and small vertebrates when they can catch them. But fruit makes up most of their diet, and in the time of year when the pickings are a little slimmer than usual, they turn to the balsa wood flowers, which provide food and a very decent energy drink in the form of nectar. Kinkajous are well built for the job: each one possesses a tongue that can stick out for 12 centimetres. The tree in flower in the time of scarcity brings the kinkajous in numbers. These flowers are an important resource, and the animals will fight for them – and become covered in pollen as they do so. They also get a liberal pollen-dusting when they merely drink from those deep white cups.

They are canopy dwellers: the trees are their home and they are well adapted for it. They even have prehensile tails, which act as a fifth limb, grasping the branches and greatly extending the animals' reach. They are one of only two carnivores (i.e. from the order Carnivora) to have prehensile tails; the other is the binturong, which occupies a not dissimilar niche in the forest of Southeast Asia, though it takes more meat in its diet.

The kinkajous' association with the balsa tree is important for both species, an example of the mutual dependence that develops between species over time. Each would find it a great deal harder to survive and make a living without the other.

But the balsa tree's strategy takes the form of an each-way bet. The kinkajous are the frontline pollinators, and they will take a heavy load of pollen for a short distance to other balsa trees nearby. The bats are less frequent visitors, but they can transport pollen from a tree to another tree that lies a fair distance away. They take a lighter load of pollen than the kinkajous but carry it a good deal further.

So, with the help of bats and kinkajous, the balsa tree thrives for its middle-distance life. Its way is one of perversity, as if it deliberately chose the options that other plants reject: not making proper wood, flowering in the drier part of the year, and being almost determinedly fragile. It is a parody of a tree and it has created for itself an ecological niche that works brilliantly.

Any strategy that allows a plant or any other living thing to become an ancestor is brilliant: that's how life operates.

The opportunistic nature of the balsa tree might, you would think, be an inconvenience to the other seedling trees all around it, all of them seeking the light with the same earnestness. You'd think the last thing a potential forest giant would need is an upstart balsa tree growing alongside. But that's not the case at all. Though the opening of the forest canopy brings light into darkness, it also creates a fierce environment: much drier than the surrounding forest, more open to wind than the usual still of deep forest, and exposed for long periods to tropical sun, which is too much of a good thing for many tender young plants. These are forbidding conditions to deal with. But a plant living alongside a balsa tree has a better chance. The faster and soon much bigger structure of the balsa protects the nearby seedlings and saplings from the worst of the exposure, offering shade, acting as a windbreak and at least some protection from the browsing mammals of the forest.

It is a phenomenon found wherever trees grow in numbers: that of nurse trees. One tree looks after a number of others, in a way that invites an observer to get sentimental about providence. The fact is that the balsa tree is pursuing its own interest, just like the kinkajous – but the potential giants exploit the swiftness of the balsa tree to further their own long-term plans.

The balsa wood tree can't survive too long: one that reaches its 30th birthday has done well. When the balsa comes down, it leaves not a patch of bare forest litter but a small and eager group of saplings. Now is the time for the giants to race for the canopy – and then to close the wound in the forest once again. Meanwhile, when the time comes, another violent storm will fell another giant, and the process starts all over again in another part of the forest.

This sequence from fallen tree to closed canopy is a succession of vegetation. The principle is not unique to tropical forests: you can find it in any environment you choose. You can leave a bowl full of water standing for a few weeks, and it will grow through a brief succession of plants until it reaches what is called the climax vegetation, which in this case, is a kind of green scum. Try the same experiment on a piece of open ground in lowland Britain and you will eventually (though this will depend on the plants' interaction with the environment and with animals it contains) get a closed canopy oak forest. When a tree falls, it triggers a succession of plants. And here in the forest of Central and South America, the pioneers like the hotlips plant get supplanted by quick-growing trees like the balsa, and climax in the closed canopy of forest giants like the Brazil nut tree, the kapok tree, and the dipterocarps with their buttress roots.

Once up there in the canopy you are in the land of leaves. Astonishing numbers of leaves, astronomical numbers of leaves. The job of a tree is to produce leaves, because leaves are life. Leaves gather light: and as we have already seen, light is life. The more leaves there are – or to put that more carefully, the greater the surface area of leaves capable of gathering light – the better it is for the plant. The leaves also take in carbon dioxide, and at the same time they give out oxygen and water; the waste products of plants that are the life-essentials of us animals. Which is rather neat.

It follows that trees are in the business of leaf production: that spreading themselves out towards the light in as wide an area as possible is the secret of success. In this way plants can again be regarded as the opposite of animals. We animals have as much of ourselves as possible inside, safe and protected. We are as compact as we can possibly can be: we need to have our organs of sense, consumption, excretion and generation on the outside; we need limbs, and if we are lucky enough to possess them, wings and tails to extend a fair way from the centre of our bodies, but in the main, we operate on the rule of maximal inside. Plants work the other way round. As you view the rainforest trees from the canopy, it is clear that being a successful plant is about maximal outside. David Attenborough stood beneath a rainforest giant that he had climbed with ropes 20 years earlier, when a mere stripling in his 70s. He recalled the great effort involved, and pointed out that a tree such as this draws two tonnes of water from root to canopy every single day.

That way you maximise your opportunities for gathering light and therefore of turning light into food by the glories of photosynthesis. The snag is that you also make yourself vulnerable. The more you have outside, the more there is available for exploitation by others. And while taking this from trees may seem to us – and actually is – the most natural thing in the world, from the tree's point of view it is a form of attack. If you eat the leaves of a plant, you reduce the plant's efficacy as an eater of light. The less a plant gets eaten, the better it will grow. So plants evolve mechanisms of defence, and so those that consume plants have to evolve ways of overcoming those

defences... and the great arms race continues, as it has continued in the rainforest for 55 million years.

Trees make their leaves tough and bitter, difficult to digest. But there are ways of getting round this. The three-toed sloths succeed in doing so by their extraordinary talent for slowness. There are four living species of three-toed sloth, all of them found in the rainforests of Central and South America. They are famously slow-moving: a speed of 0.15 mph has been suggested. That seems amusing to those of us who live at a different pace, but what matters is that, for the sloth, slowness is not a terrible mistake, but an extremely efficient way of living.

The slow place allows them to hold a great deal of chewed leaves in their stomach and gut at the same time, and to let it digest at a leisurely pace. The stomach contains four chambers of digestion, so that the sloth can maximise the nutrition from the leaves it eats. This system wouldn't work if the sloth was tearing about and burning up energy, but in the sloth's stillness is the answer: if you conserve energy by being still, you need less fuel. A sloth will sleep for 15 or even 20 hours in a day. A sloth can hang from the branches in its favourite position, with a bellyful of leaves, digesting away at the most leisurely pace imaginable. The other organs don't get crushed by the weight of the belly (which happens with a morbidly obese human), even though a sloth's full belly can be a third of the animal's total weight. That's because stomach, lungs, kidneys and other organs are attached to the lower ribs or the pelvic girdle.

It can take a full week for a leaf to pass through a sloth. The process is helped along by the heat of the sun: a sloth will bask, belly up, at the top of the canopy and the sun warms the microbes and bacteria in the sloth's digestive system and helps it to hurry – insofar as such a term can be used of sloth – along. A sloth works on the opposite principle to other mammals: it east less when it is cold, more when it is hot; in fact when the weather gets cold, by rainforest standards, it is possible for a sloth to die of starvation with a belly full of food.

The lifestyle of sloths is strange to us; so much so that we have named them

OPPOSITE
Slower and slower: brown-throated three-toed sloths – mother and young – in Tenorio Volcano National Park, Costa Rica.

after one of the seven deadly sins. But their life in slow motion allows them to live for up to 30 years, comfortably exploiting the forest and the best defence mechanisms the tree can throw at them.

Leaf-cutter ants operate a different strategy, one that is, at least in part, based on speed. They have jaws that open and close 1,000 times a second, and that allows them to operate with neat surgical swiftness. That is not only highly efficient as a mechanism for cutting leaves, it also works well because the tree is unaware of what is going on.

Perhaps it seems a step too far to start talking about the awareness of a tree: one step from tree-huggery and giving each tree a pet name. But if a living thing responds to different stimuli, we have to presume an awareness of some kind, while being very careful not to make any assumptions about what kind of awareness that is. Your favourite chestnut tree in the park is unlikely to be aware of your affection, but it is certainly aware when it is autumn, because it starts shedding leaves. It is equally aware of spring, because it starts growing them again. And a tree will also be aware – without having a central nervous system or the power of reasoning – that it is under attack.

It is all about the basic economics of staying alive. If a herd of wildebeest ran like hell every time they saw a lion, they would never stop running and never have time to eat. So they adopt a strategy that allows them to eat while keeping lions in view at a safe(ish) distance. A tree can defend itself by pushing toxins out into its leaves when they are being eaten. But this process is (like running away all the time) expensive of the tree's resources. If they reacted every time they felt the tiniest nibble, they would spend the whole time creating toxins rather than reserving their energies for new growth. So they wait until a leaf receives damage of 20 per cent before acting.

The leaf-cutter ants evolved the most elegant response to this strategy: they remove, say, 19 per cent of the leaf and then leave it alone. Again, we must be wary of attributing thought and calculation into all this: but the ants consistently operate by inflicting a level of damage below that at which the tree starts to defend itself. You could say, if you like, that the ants are fooling the tree. Or you could say that the tree is successful because it habitually sustains a level of damage that it can cope with: after all, even a damaged leaf is still photosynthesising at a level of efficiency greater than 80 per cent.

But rather than attributing random personalities to the exchange, let us appreciate that this is a balance that works. Without benign feelings for each other, tree and ant live together and both continue to function: the ants are getting what they need without compromising the tree's future to any disastrous extent.

There remains the problem of the toughness of the leaves. The ants' solution to this is staggering, and it requires a complexity of lifestyle that is startling in its complexity. The great evolutionary biologist Edward O Wilson, also a specialist on ants, said that leaf-cutter ants are 'the most complex social creatures other than humans' on Earth. That's quite a claim, so let us look at the ants more closely.

There are about 40 species of leaf-cutter ants in the Central and South American tropics, but they all live in roughly the same way. They live in colonies that can hold as many as eight million individuals, and they construct

ABOVE
Heavy laden: a worker leaf-cutter ant takes another monster fragment back to the nest.

nests big enough to manage these colossal numbers: these can cover an area as large as a tennis court and go as deep as a two-storey house. The colonies are centred on a single queen, mother of all she surveys, who is capable of laying 30,000 eggs in a single day. She can live up to 12 years and never once leaves the nest. She will lay 200 million eggs in the course of a lifetime.

The leaf-gathering strategy of the leaf-cutter ants is well known: they carry pieces of leaf, walking in processions like demonstrators with a series of banners. They are an image of toil, like Boxer in *Animal Farm*; it's as if each individual had vowed: I will work harder. They are famously capable of carrying their own bodyweight: the heroic drudge taking on a series of impossible tasks and always succeeding.

This is all fascinating enough: the endless procession of ants, each with her trophy – these working ants are all females – which goes back to the nest to nourish the colony underground. But not a single member of the colony eats those leaves. They can't. The leaves are far too tough; no ant could possibly digest them. Instead they use the leaves as compost. They make gardens deep in the darkness of their hive, and these gardens produce fungus – which is rich in protein and wonderfully easy to digest. The food from these gardens feeds the larvae that live within the colony and which are the future of the hive.

This system is so complex it can only work on the principle of the division of labour. Each ant in the colony is a specialist: there is no room here for the polymath: the renaissance ant is not wanted in the society of the leaf-cutters. So let us start with the foragers, who are the visible ones, the ants we are familiar with. These explore the area around the nest for the right leaves, non-toxic and suitable for the underground gardens. They return to the nest with their banners of leaf, and their success attracts more ants. They leave a trail of chemicals that other ants can now follow. That way a good resource will be fully exploited.

Often these familiar processions of leaf-bearing ants will be attended by guards. There are species of parasitic flies that seek to lay their eggs in the heads of foraging ants, and when these hatch, they burrow into the ant and kill it. The guards defend the foragers from this danger, and on the homeward leg will often hitch a ride on the piece of leaf, adding to the workload of the individual they are protecting. Some species send their larger and more efficient foragers out at night, when there are no flies about. The foragers they put out on the day-shift are too small to be exploited by the flies.

Other workers within the colony do the many difficult and complex jobs underground. The gardeners tend the fungus gardens and also act as nurses, caring for the eggs, larvae and pupae – collectively known as the brood; the brood is generally located in the tunnels and chambers round the garden.

Other ants act as excavators, constantly making and maintaining tunnels and chambers throughout the sprawling colony. They will create chambers for the disposal of rubbish. It is obvious that so large a population living so closely together is vulnerable to disease, and it follows that nest hygiene is of critical importance to the colony. Within the colony there are ants whose sole job is rubbish disposal. This segregation minimises the possibilities of their introducing disease to the colony from the dirty and essential job they perform. They avoid contact with the fungal gardens, with the nurses and with the brood, and of course, with the queen – after all, an infected queen is the death of the colony. So by creating this caste of untouchables the ant colony keeps itself fit and functioning.

Finally there are soldier ants, the largest of the all the worker-ants in the colony, and their job is to guard the entrances to the nest and the trail paths. All this degree of specialisation is unique to leaf-cutter ants. There are other species of ant that go in for fungus gardening, but with these, the tasks are interchangeable. The leaf-cutters take this up to a higher level of organisation.

But the entire system is dependent on the least obviously spectacular part of this system: the fungi. These are the greatest and most efficient herbivores in the forest, and they achieve this by exploiting the ants. A big colony carries 50,000 pieces of leaf into the nest every day. By kindly permitting the ants to labour for them, to gather food for them, already carefully cut up for their convenience and to protect them, this vast and complex society has evolved for the glory of fungi.

The details of the lives of leaf-cutter ants show just how complicated life in the rainforests has to be. There are no simple solutions. The complex nature of dependencies and interdependencies is an inevitable part of the long-term stability of the rainforest: precisely, of course, what is under threat across the world. It is all connected to the fight for light: everything in the forest starts with light, as it does everywhere else. But just as the leaf-cutter ants find a way of exploiting the system in a manner that looks, to human

OVERLEAF
Working party: leaf-cutter ants live in communities of up to six million individuals, a population dedicated to agriculture.

eyes, a little bit like cheating, so the world's largest flower breaks the rule that plants must have light.

Rafflesia, named for Sir Stamford Raffles, who founded Singapore as a British colony in 1819, is a plant that doesn't bother to photosynthesise. That doesn't contradict anything that was said earlier: the plant generously allows other plants to carry out the work of processing light into food – and then takes advantage of them. In other words, it's a parasite. There are 28 species of *Rafflesia*, all from Southeast Asia, and they make their living by absorbing the energies of trees and jungle vines. These vines are themselves parasites – structural parasites, that is. They take nothing from the trees they climb on but their support. They use the trees to reach the light, and they can grow to a length of more than a kilometre as they search for it. They clamber to the canopy where they can feast on the brilliant light available: when you look on the jungle from above, 60 per cent of the leaves you see are not on the trees but on the vines they support.

Rafflesia grows on these vines, so it is a parasite on a parasite. All you can

ABOVE AND OPPOSITE
Nothing but flower: the giant Rafflesia blooming on the forest floor in the Maliau Basin, Malaysian Borneo.

see of them, usually, is the flower: and it really is a monster. Some species will routinely produce a flower that measures one metre, in diameter, and weighs 10 kilograms. The record is 1.2 metres: not the flower to choose for your button-hole. (The titan lily, one of the stars of *The Private Lives of Plants*, a previous BBC series with David Attenborough, is still bigger, but it's not a single flower: it's technically an unbranched inflorescence, something that includes many flowers, though it looks like just one to most of us.)

Rafflesia are not the most attractive flowers, at any rate not to humans: they look and smell like rotting meat. What they attract is flies and beetles, which can be found in numbers at the ground level in the forest, and they do the work of pollination as they search in vain for a cache of meat within the giant flower. The rainforest throws up oddities again and again: sometimes subtle and tiny and invisible to most of us; sometimes in gigantic and spectacular form. In the 1960s Rafflesia was the first thing you saw when you entered the Natural History Museum in London: a perfect introduction to one of the world's greatest treasure-houses.

Perhaps the true stars of any television series about plants are in fact fungi, not plants at all – for without them the plants would find it hard to survive and prosper in their present shape. But perhaps these fungi are more like the people whose names you find in the long list of credits at the end: people who hardly ever, if at all, appear on the screen, performing functions obscure to most viewers but are essential to the finished product. When we hear the term fungi, we generally understand it to mean the fruiting bodies of fungi: mushrooms and toadstools and so forth. But that's not a helpful idea: it's like confusing an acorn with an oak tree. Most of the life of fungi is lived as slim white threads that creep and divide through the soil. Like us animals, fungi don't make their own food, like us they are pure consumers. But the fungi's process of pursuing their own ends helps plants to prosper. In the rainforest the fungi feed on the dropped plant and animal matter that covers the forest floor: and by doing so they make the nutrients they absorb available to growing plants. Fungi unlock the resources of the forest and make them available – to the forest, in the glorious circular process that makes the rainforest what it is. Much of this process benefits both parties, one of those mutualistic relationships that give such satisfaction to the human observer. Not all natural relationships operate like this, as we shall see shortly.

OPPOSITE
Fungi live mostly out of sight, but they play an essential part in the ecology of every rainforest, making such extravagance possible.

TOP *Cup fungus from the Danum Valley, Malaysian Borneo.*

MIDDLE LEFT *The bioluminescence of some fungus is called fairy fire or fox fire. In some parts of Africa it is called chimpanzee fire.*

MIDDLE RIGHT *The same fungus in daylight.*

BOTTOM *Fungal threads – mycelium – grow into and consume a decaying leaf on the Osa Peninsula, Costa Rica.*

But in the forest, the fungal threads – mycelium – act as a kind of secondary root system: the fungus benefitting from carbohydrates produced by the plants, the plants benefiting from the much greater absorption of water and nutrients from the mycelium. The network links one tree with another across the forest: a tree under attack can inform other trees by passing a chemical message that travels through the network of fungi, which has been called the wood-wide web.

In rainforest some species of fungi put out fruiting bodies that glow in the dark, looking like the decoration for a ghost train. No one is sure what the fungi gains from this display. Sometimes these displays of light – of bioluminescence – can light up a tree. In Costa Rica, a single tree was filmed glowing as if lit within: but in fact the glow came from the branches – hyphae – of the mycelium. The reason for this remains baffling and obscure: but the rainforest is a constant provider of mysteries. A tree that glows in the dark is just one more layer of complexity to the endless complexities of tropical forest.

The wet tropics are invariably filled with life, in a way that is beyond the easy imaginings of those of us who live in temperate zones. The complexities continue even when the ground rises up to chillier altitudes and the nature of the vegetation is forced to change. Mount Kinabalu in Borneo reaches up beyond 4,000 metres, and the classic rainforest at the foot changes step by step in a series of zones as you climb towards the summit; this is more of a demanding walk than a feat of alpinism. As you leave the lowland forest behind, you move into the lower montane zone, characterised by species of oak and chestnut, and then into the upper montane cloud forest, full of mosses and liverworts, and then up towards the subalpine meadow zone and the summit. The higher up the mountain you go, the harder it becomes for the plants you observe to get the nutrients that are so readily available in the forest at the bottom – under the rule that the forest feeds the forest. Where the forest changes as conditions become more demanding higher up the slope, the system becomes less rich. One solution to this shortage of nutrients, one that has fascinated people from Charles Darwin to the makers of horror films, is to become carnivorous. There are plants on Mount Kinabalu that get what they need by trapping insects and consuming them.

Ten different species of pitcher plants grow in the two montane zones of Mount Kinabalu, tempting in insects with an offering of nectar on the rim of

the pitcher. The insects come in to drink and then find that they are on a
slippery slope. They can't get a grip on the rim of the pitcher and so they
slither to their doom, for the pitcher contains not more nectar, but a liquid
full of digestive enzymes. But one plant has come up with a refinement on
this that deliberately lets the nectar-drinker get away.

It begins as an insectivorous pitcher plant, functioning in the usual way.
It is a meagre and precarious existence: but the plants that manage to reach
maturity then change their shape and their way of life. They now have a
different kind of pitcher and they use it in a different kind of way. The
pitchers grow larger, reinforced with lignin, the woody substance that makes
plants robust. This has an angled lid that is not at all inviting to insects. But
the shape of the plants and the treasure of the nectar inside – this species

ABOVE
*A mountain tree shrew feeds on the nectar of a pitcher plant on
the slopes of Mount Kinabalu, Malaysian Borneo.*

produces more nectar than other pitcher plants around it – is extremely enticing to small mammals, tree shrews in particular.

It was once thought that the pitcher plant's strategy was to catch and consume the shrews, but clever positioning of camera-traps revealed the truth. The shrews come in all right, and they enthusiastically drink the nectar they find under the lid. When they do so, they find that the plant is so cunningly constructed that a shrew must hang its back end over the middle of the pitcher. As the shrew drinks it will often defecate – leaving a good nitrogen-rich source of nutrient right where the plant can use it.

This is not a system based on the blind hope that the shrew will perform when required. For a start, the metabolism of all tree shrews is fast. They depend on a rapid throughput: one end to the other in 15 minutes in some species, no more than an hour in the slowest. That puts the odds in the plant's favour at once. But the plant may take this a step further. The nectar that rewards the shrew for its visit contains a laxative, so that the shrew can't help but oblige the plant it is visiting. The plant is manipulating the shrew's digestion for its own purposes.

There are times when mutualism of this sort gives the pleasing illusion of the benignity of nature. The truth is that nature is not benign any more than it is vicious. No moral term is appropriate. Nature just is: organisms doing the best they can to make a living and to become an ancestor. And certainly, what Darwin called 'the struggle for existence' can be experienced with immense vividness in a tropical forest, perhaps more vividly than anywhere else on Earth. The inextricable nature of life and death is brought to our notice with an intensity you can avoid in a well-managed English garden – so long as you don't look too closely. But what happens in subtle and elusive fashion in some environments is there right in your face in others.

Much of Australia lies in the tropics; the Tropic of Capricorn goes right through the middle of the great island-continent. And in the most northerly parts of the island-continent, at the tip of Queensland, you can find tropical forest and there you can examine the struggle for existence with a rather disturbing intimacy. In any rainforest, the usual pattern is for a closed canopy of trees of uniform height, punctuated by still greater giants that raise their head above the rest. These are called, unromantically but accurately, emergents. The poison arrow tree is a classic example of this: and it is

OPPOSITE
Swapping food for shelter: a Hardwicke's woolly bat weighs only four grams and can shelter during the day inside a pitcher plant. The bat gets shelter; the plant gets nutrients from the bat's droppings.

particularly attractive to a species of bird called metallic starlings; these are birds that look dull in some lights but gorgeously iridescent in others. The poison arrow tree is a widely spread group, and their toxic nature has been exploited for darts and arrows in some cultures.

Metallic starlings migrate between New Guinea and Australia, and do their breeding in Australia. They like to do so in big colonies and for this they select a poison arrow tree, standing proud of the forest; trees with a thousand nests are not unusual. The tree's eminence is only one of the tree's advantages: the better part comes in the smoothness of the bark. That makes the tree very hard for snakes to climb: snakes push against their reverse-pointing scales on their bodies to gain purchase for forward movement: no grip, no movement.

A nesting colony of birds sounds like the best place in the world to celebrate the kindness of nature: but it is the struggle for existence that is more obvious. The stench of the droppings of thousands of birds is always the first thing a human visitor notices. The second thing you notice is that, to many creatures, thousands of baby birds represent a bonanza: an explosion of life

ABOVE AND OPPOSITE
A strange togetherness: metallic starlings and the poison arrow
tree in northern Queensland, Australia.

that must be exploited, as you would a tree full of ripe fruit. So the poison arrow tree with a nesting colony of metallic starlings operates an immense pull on the animal life of the ecosystem. This explosion of life is also an explosion of death: eggs and nests fall to the ground, and frogs, toads, snakes, birds of prey and dingoes come in to feed beneath the tree: and so too do many invertebrates, centipedes and flying insects. The density of animal species around a poison arrow tree with such a nesting colony is 100 times greater, and sometime up to 1,000 times greater than it is elsewhere in the forest.

Even when the starlings have gone, the pattern of abundance continues: the richly fertilised ground allows fallen seeds from the forest to germinate and grow fast, bringing in the herbivores like cockatoos, brush turkeys and pigs. But the annual rhythm of bonanza can't last forever. Eventually – within as little as 15 years of service as a nesting tree – the accumulated droppings will kill the tree. It will fall, leave a gap and create another form of opportunity. Life and death, death and life, inextricable one from the other, both aspects of the same process.

The great trees like the poison arrow tree, the emergent trees, are the stars of the rainforest. There are great privileges that come with this status, standing proud above the rest, and the greatest of these is access to as much light as they could possibly want. Once they have risen above the competition they can bask in the sunlight and their own glory: able to grow taller because they are taller. They can make more food for themselves precisely because they have succeeded in making so much food for themselves before.

Down below them in the canopy the situation is more fraught. Each tree is competing with its neighbours: down at the bottom they compete for access to moisture and minerals, and up at the top they compete for light. But it's no good crowding into each other like a bully trying to get served in a busy pub: once the trees start touching and overlapping they are doing themselves no good. They are forcing their own leaves into the shade of a neighbour: wilfully depriving themselves of light. That's not going to help a tree at all: it's an inefficient use of resources. So they have created a kind of tradition of respect: one that is based on each tree's own best interests. The trees of the canopy each draw back from another, creating and therefore permitting a personal space for each tree. It's called crown shyness and it is, as you would expect, all about light. It is the light from the red end of the

OPPOSITE
The buttressed giant: a dipterocarp in the Danum Valley, Borneo.

spectrum that does most to fuel the growth of plants, but once the leaves are shaded they receive only red-depleted light. They can sense this alteration in the quality of light and, from this, infer the presence of a neighbour. So each tree allows the other a kind of social distancing, something that benefits both plants, and saves them both from futile effort.

But the emergents don't have to worry about neighbours. In the rainforests of Southeast Asia, 80 to 90 per cent of the emergents belong to the family of dipterocarps, named for their winged seeds. These are the trees that we think of when we hear the word rainforest: immense trunks supported by giant buttresses: roots that start a good way up from the trunk and stretch out to keep the giant balanced – and so able to grow taller than all the rest.

To stand at the base of such a tree is a powerful experience: one that gives you a crick in the neck and an instant naïve question. How come this tree is so damn good at growing? The fact is that dipterocarps – there are nearly 700 species – really are capable of producing more wood than their neighbours. They are never pioneers, in the manner of the balsa tree that grows on the other side of the world, but they still grow fast. They also live a long time: there is nothing here-today-gone-tomorrow about them. It's possible this

ABOVE
Built to last: the big buttress roots give a certain stability
to many rainforest trees; this one in French Guiana.

advantage comes from a more intimate relationship with the mycorrhizal fungi, the threads of mycelium that live in the soil and all around the roots of the trees. The fungi they associate with form a sheath over the outside of the roots of the tree, a difference that may lead to more efficient gathering of the nutrient resources that the fungi bring in their gift.

A massive and unmissable tree like an emergent dipterocarp is always going to be the centre of attention. When it produces its fruit, which contain the seeds, no one is going to miss the fact. The tree's emergent status is a come-and-get-me signal to all the many forest creatures that feed on seeds. The tree is made vulnerable by its own magnificence.

The dipterocarps have a strategy to counter this problem: and it is to become even more prominent than usual. Every seven years they produce a great bonanza of fruit – and so does every other related tree. They synchronise their fruiting. The result is that every few years the forest is full of dipterocarp seeds: so many that no matter how hard they try, the seed-eaters of the forest will never be able to eat them all. This is called a mast year: we see the same phenomenon, though to a lesser extent, in oak trees in colder climates. The strategy of superabundance is about predator satiation: by producing far, far

ABOVE
Vigorous foragers: bearded pigs on the forest floor, Malaysian Borneo.

too much the dipterocarps always have enough. There is a payback, of course, as there is to all strategies: the tree might die in the long gap between mast years. But here is an answer that works well enough – though it works only if you are mighty enough and numerous enough to satiate your predators.

As a result of this policy of excess, many of the seeds will reach the forest floor and germinate. Many fall all around the trees, not traveling far, despite the wings on their seeds. Many will be eaten by bearded pigs, who love a mast year; the tree is so important to them that they time their breeding to the years of dipterocarp excess and breed every seven years. These pigs will eat and move away, later inevitably defecating. Undamaged seeds in their droppings will germinate, and grow with the help of the generous dab of fertiliser.

The threat of death within the seven-year cycle of masting is a real one. When you stand head and shoulders above the rest, you have all the advantages, but by doing so you make yourself vulnerable. The vulnerability comes not only from the height above the rest, but also from the moisture that the trees release into the air. It's been found that a disproportionate amount of water vapour gets into the system from the emergent trees: relatively few emergents do more of the meaningful transpiration of water than the lowlier trees in the canopy. This, as we have seen, creates rainfall – for it is the rainforest itself that bring the rain to the forest – and with them, the lightning storms that add such drama to forest life. The emergents are the prime targets for lightning, which loves to find the quickest way down. It often happens that for the greatest trees of all, in every rainforest system on Earth, that the triumph of the emergent is also its doom: that victory is the beginning of the end.

These emergents are of course the most tempting of all trees in the wood to human loggers. This creates problems for the rest of the forest that survives. It is a double process of drying out. First, the trees that contribute the most to the cycle of rainforest rains are taken out: and second, the gap they leave causes the forest all around to dry up as well. It's a double-whammy.

We have a natural tendency to misunderstand forest. It seems simple: a lot of trees, and a lot of animals that live in, around and beneath them. We can't get our heads around the complexity of it. A computer with a game of patience or solitaire up on the screen looks simple too, but of course it isn't. And if we spill a glass of water in the wrong place the computer is in deep trouble.

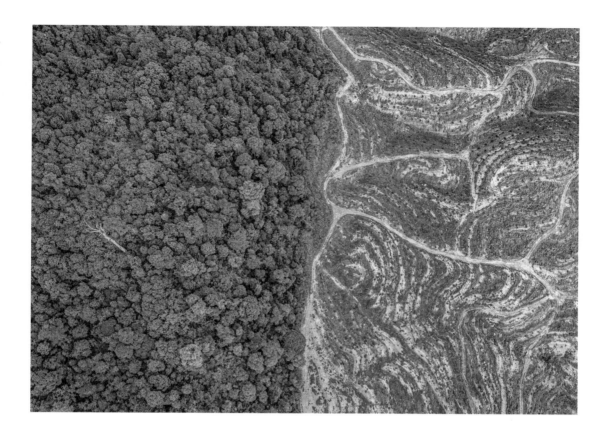

That is what has happened to rainforest. By thinking it's all a pretty simple matter, we fail to do justice to its complexity. The problem is that leaving token chunks of forest here and there is not the same thing as having forest. We have seen the way that forest repairs itself: a giant falls, the sprinters have their day, the middle-distance runners like the balsa tree flourish briefly and purposefully, then the giants take over and the canopy is repaired: so the forest that looks so still and permanent is a dynamic, happening sort of place, a sort of living perpetual motion machine that is fuelled by itself and by the light of the sun. A tree falls, but that's not a disaster, it's part of process. The forest takes the wound and the forest heals itself, a little in the way that our own bodies recover from hurt.

ABOVE
Palm oil is an ingredient in thousands of useful commodities.
Its cultivation has taken up half of Borneo's rainforest.

But when the forests are fragmented and broken, that process no longer takes place. The edges are the vulnerable places, and when a forest becomes all edges, it becomes a thing of ever-multiplying vulnerabilities. That has given rise to one of the most bizarre and macabre problems of deforestation ever identified. It is the problem of the silverleaf *Desmodium*: the sticky pea of Madagascar. It is as if a conservation issue had been visited on the world by Batman's most fearsome enemy, the Joker.

It's all about an invasive plant; there will be more on this challenging question in the last chapter of this book. Here we are dealing with a plant that is native to Central and South America and has been domesticated as a protein-rich fodder crop. It is classified as a legume, like beans and peas, and it grows well in tropical places, while having the additional strength of being cold-tolerant and even frost-tolerant. It tends to be one of the first plants of the year to show. It is classically used as rotational grazing for livestock, and it can also be cut for hay. Well-managed, it is a pretty good thing in the right climate.

OPPOSITE

TOP *Cutting edge: Borneo rainforest now shares a boundary with oil palm plantations.*

MIDDLE *Another one gone: logging in Southeast Cameroon.*

BOTTOM *A beast with troubles: the radiated tortoise of Madagascar is hunted for meat and for the exotic pet trade – and their forest habitat is being destroyed.*

ABOVE

Once it was all jungle: Madagascar has 10 per cent of its forest left.

But, in Madagascar, *Desmodium* has gone feral. In places it now dominates the native vegetation. It is a trailing perennial, and can grow several metres in length, clambering over native plants and outcompeting them for light. It spreads easily: the hooked seeds evolved to stick onto the fur of passing mammals, but they also readily cling to the clothes of passing humans. The seeds spread easily and germinate readily, and in this warm, wet place there is no cold season forcing them to die back. Everything is in their favour.

The plants have a defence mechanism: their hairy stems deter potential predators from climbing aboard. In this species, the hairs are hooked: and that has created an extraordinary problem. In Madagascar, the combination of high humidity, warmth and the plant's ubiquity had turned *Desmodium* into

PREVIOUS
When the pygmy elephants of Borneo get in the way of oil palm cultivation, there is only one winner.

ABOVE
Trapped: the hooked hairs on the stem of silverleaf Desmodium have snared a dragonfly in the Alaotra-Mangoro Region of Madagascar.

a plant of death. That is not being overdramatic: frogs, chameleons, bats, insects and many other different species reach the plants and get stuck on the hairs, often at night. Once there, they have nothing to do but die. They are exposed to the sun and there they dry out: preserved in their last agonies, trophies of the invading *Desmodium*. It is like a visit to a plague village: unaccountable death in numbers that are hard to take in. Researchers for the televised version of *The Green Planet* talked about a tragic scene, about a forest of death. Something about this strange phenomenon hits us all on a very deep level. The problem was largely unknown before *The Green Planet*, and was a revelation when programme researchers took their findings to the Royal Botanic Gardens at Kew in London. The problem is doubly disturbing

ABOVE
For over 200 years ranchers have been cutting down trees to create grazing land for cattle; an example in Costa Rica.

because of its association with rainforest. All those who visit rainforest come away with a head spinning with the wonder of it: a rainforest is, as David Attenborough has spent a lifetime pointing out, teeming with life. It is the genius of humankind to create from rainforest a place that is teeming with death.

Perhaps the saddest thing of all is that we can no longer talk about the wonders of the rainforest without talking about destruction. Humans cause gaps in the forest at a much faster rate than the natural processes. The edges are always the most vulnerable parts of any forest, and the process of human destruction creates a series of forest-edges. These dry out, and are more vulnerable to fire. Forests are reduced to isolated clumps of trees: often these contain small populations of animals that can travel nowhere else and can only breed with each other. This creates inbreeding depression and an inevitable loss of resilience. The fact is that clumps don't operate in the way that entire forests do. It's been calculated that a forest can only function as the vast, complex and thrilling environment we have been celebrating if it can maintain what is called a minimum critical size. A large forest is not uniform but variable, and so full of different ecological niches, that is to say opportunities for organisms to make a living. The more niches you have, the more species you will get, especially in a long-term stable environment. A small forest must therefore hold fewer species. Many large vertebrates need a wide home range: they can't function in a smaller space. If you reduce the size of forest you are likely to wipe out the major predators. That means the herbivores thrive in unnatural profusion. There are more animals out there eating plants, and that puts stress on the plants that provide their nutrition. An open forest with its many edges is full of light, and therefore the shade-loving plants die. The species that loves gaps – those that come in when a giant falls – start to dominate. A small rainforest is not really a rainforest at all, not in the sense that we understand the term: the richest and most variable species-rich places on the surface of this planet.

Let's call that the heresy of tokenism, or the Panda Fallacy: the idea that we can protect biodiversity by making sure that we hang on to a token amount of each species, especially the glamourous ones. The rainforest tells us that what matters, in terms of diversity, is a dynamic environment packed with species, one that takes on carbon, gives out oxygen, transpires water, and effectivity acts as the planet's air-conditioning system: precisely what we need to slow down the disaster of climate change.

In the televised version of *The Green Planet*, David Attenborough was in Costa Rica, standing in exactly the same place he had stood 30 years earlier. Back then it was all grass and cattle; now, in a regeneration project, there was new forest striking up again. It was a glorious sight: a bright spark of hope to light up the gloom. He read from the diaries of Charles Darwin: 'Among the scenes which are deeply impressed on my mind, none exceed in sublimity the primeval forests undefaced by the hand of man.' Attenborough then added: 'He would struggle to find such a place today.' He then dealt out a killer stat: 70 per cent of all rainforest trees now grow within 100 metres of a manmade track or clearing.

We have become used to the association of rainforest with sadness and even despair. But the process of forest destruction can be stopped, and in many places it is being put into reverse. We can find just such a thing going in Costa Rica, where the government, rather than private and charitable organisations, has taken the lead. They are prepared to pay landowners for environmental services: that is to say, for not cutting down trees, for allowing the forest to regenerate. In 70 years, Costa Rica lost 80 per cent of its forest: but since 1996 the process of healing has been established in many places throughout the country. It is possible to stand on land that had been cleared for pasture – pasture to feed cattle and the human passion for beef – and to be unaware of that fact. The forest is back. Not entire and not the same, but like a patient on a sickbed, the forest is responding to treatment, sitting up and taking nourishment. It can be done, and in some places, it actually is being done.

David Attenborough, visiting the healing forests of Costa Rica, said: 'We have the ability to repair considerable damage to our rainforests. It will take the cooperation of nations around the world, but it is the only way in which we will be able to preserve the treasures of the tropical rainforest for future generations.'

OVERLEAF
David Attenborough with the Usain Bolt of the rainforest: a young balsa tree at La Selva Biological station, Costa Rica.

INSET *Ascent to a different world: David Attenborough rides the canopy tram in the rainforest of Costa Rica.*

DESERT
WORLDS

We call the dry places of the world deserts because they are deserted. A desert island is not an island covered in sand and cacti, it's an island with nobody on it. But the reason that a desert is deserted is because there is no fresh water. And where there is no water there is no life.

But here we can find one of the routine miracles of the natural world: there is life in the desert after all, and that is because there is water in the desert after all. Just not very much of it – but then you don't need much water at all for life to take place. The great moist rainforests are teeming with life, in numbers beyond easy human imaginings. But the great dry places of the world also have life – not exactly teeming, but certainly out there and hanging tough, doing that life-loving thing of surviving and trying to become an ancestor.

Plants need water. We all know that. We've all seen a neglected houseplant droop, almost audibly begging for water. The idea that plants need water is part of the bleeding obvious, as Basil Fawlty would say. So how do you answer the intelligent four-year-old who asks why? Why do plants need water?

There are quite a lot of answers to this question, and each one of them is not only right but essential to survival. A plant needs water just to be. That's because a plant is mostly water: say 90 per cent. If you're an adult human reading this book, you are about 60 per cent water. We all need water to survive, but in some ways a plant needs water even more than we do, which makes the existence of desert plants even more astonishing. A plant deprived of water wilts. That's because there is a loss of water pressure within the cells. The cells become less turgid and so the plant begins to collapse and to lose leaves. A plant needs water to stay upright.

Seeds need water before they can germinate. Contact with water activates the enzymes within the seed, and also swells and softens the hard shell around the seeds, allowing the root to break out, and then to dig down in search of water. After that, the green shoot develops, and it's able to stand up because of the water that comes into the plant via the root. Plants also need water for photosynthesis to take place: water vapour out, carbon dioxide in, sunlight providing the energy that drives the process. Water is driven through plants by the process of transpiration, as we saw in the last chapter: this allows the plant to take on carbon dioxide, to cool, and to bring nutrients from the roots through the entire plant.

It follows, then, that most plants are unable to survive extended periods without water. Those that can cope with such deprivation are among the most

OPPOSITE
Deserted: a long-ago flash flood has left its traces in the cracked clay; behind it creosote bushes hold on to life in the Mosquito Flats, Death Valley National Park, California, USA.

remarkable living things on the planet. They seem almost a contradiction: life in a place that can't possibly sustain life. But there is a little water in the desert – not often very much, and it doesn't get there very often, but there is water there all right – and in such a world it is the most precious thing of all. The idea that life can exist in such forbidding circumstances is bewildering, and to us humans really rather moving. The determination of life to make life, no matter how difficult it all is, has driven the world ever since the first spark of life came into existence. That is the principle that made desert plants – and also made us.

Plants in a desert must find a different way to live. They must live and grow and reproduce at a pace dictated by the availability of water. For some, that requires a drastic slowing-down: living at a speed that makes a three-toed sloth look like Usain Bolt. In the course of filming the televised version of *The Green Planet*, David Attenborough visited a plant in the Mojave Desert. This is

the driest place in North America, and it lies in California and Nevada. Here, he visited a creosote bush. It doesn't look much, it has to be said, but it's reckoned to be one of the oldest plants on Earth at a little under 12,000 years old. He stood in the exact same spot that he had done 40 years earlier, and noted that the plant didn't look very much different. It had grown a little since his previous visit – and the total growth was measured at 12mm. Say the width of your index finger.

Let's savour those 40 years for a moment. A human born on the day of Attenborough's first visit to the plant could, by the time of Attenborough's second visit and with enough enthusiasm from all

ABOVE
An intimidating environment: the sun strikes a dragon's blood tree in mountainous desert, Socotra, Yemen.

concerned, be a grandparent. In the same time as that creosote bush grew its dozen millimetres, a balsa tree could have sprouted from seed, grown 30 metres to the canopy, put up its flowers, had them pollinated by eager kinkajous, put out fruits and seeds of its own, collapsed and died. But out there in the desert, getting minute quantities of water from the sand, the creosote bush has made its tiny advance: a glorious triumph of not dying in a place where dying is the easiest thing in the world.

A creosote bush will seldom grow to more than three metres in height. They have the knack of splitting into separate crowns, cloning more bushes from themselves: genetically identical, essentially the same plant, making a modest circle as the clone colony expands. The so-called King Clone in the Mojave makes a rough circle of about 20 metres, and you wouldn't know you were in the presence of one of the great natural wonders on Earth if you hadn't been told. A creosote bush can survive with no water at all for two years; the strategy of waiting is essential to many of the plants that make a go

ABOVE
David Attenborough and the King Clone creosote bush,
which is 11,700 years old.

of life in the desert. Eventually, some day, some moisture will come, some rain will fall. The advantage is to the plants than can do the waiting without dying in the process.

It is a question of maximising the resources available: that's true of deserts, and it's true of all other environments. It's just that with deserts the principle is more dramatic – and it's the principle behind the stands of poplar trees in the Taklamakan Desert in Northwest China. This is the second largest shifting sand desert (and the 16th largest desert) in the world, with a temperature range that can drop to below minus 20°C and rise beyond 40. But the desert has a strange river, the Tarim, which carries in water from elsewhere and dries out before it reaches the sea. The seasonal rise and fall of this river create an opportunity: the trees have a wide, spreading network of

ABOVE
The Taklamakan Desert is the largest desert in China and lies
in the northwest. Here the desert poplars find a way to live.

roots that allow them to take in water very rapidly when it's available, and so to power a brief period of fast growth. It is this ability to change gear that is behind the success of these poplars: they can grow at a pace that is very close to stop, and then react with alacrity when there is water available. The delightful incongruity of the desert poplars has made them a tourist attraction, one that is now threatened by overenthusiastic water abstraction from the Tarim River.

It is wrong to think of deserts as places with no water. It's just one more occasion when our human perceptions find it hard to cope with an alteration of scale. There are times when desert can get quite wet; it's just that they are not very frequent and they don't last terribly long. Water is not something you can rely on in quite the same way as you can during an English winter.

Water will come to a desert, sure, but you never know quite when. And when it does come, you never know how much you will get. The plants that survive are those capable of making the most of the opportunity presented. That is of course true of every other environment: but desert plants can be defined by their ability to respond to water – rapidly and decisively – on the occasions when it is presented to them.

I once had the bewildering experience of the Kalahari in flood: a desert full of ducks and flamingos, patrolled by African fish eagles that had arrived in numbers to feed on the frogs that were emerging from the dry soil in the onrushing rain. I will therefore always think of the Kalahari as a green place where the living is easy, and it's a valid enough impression, so long as you remember that you only get conditions like this every ten years or so.

Conditions can be still more extreme in other deserts across the world. The heart of the Australian continent is dominated by Lake Eyre, more correctly taking in the aboriginal name and called Kati Thanda-Lake Eyre. And under normal circumstances, the lake is not a lake as we normally understand, for it has no water. It is just more desert. It is at the heart of a river system that drains into land rather than the sea, like the Tarim River with the poplars that we have just discussed. Lake Eyre is one of the largest such systems in the world. It covers 1.2 million square kilometres of South Australia, Northern Territory and Queensland; on a map it looks so big there doesn't seem to be much of Australia left outside it. It includes the lowest point of Australia, at 15.2 metres below sea level.

And it is normally a salt pan: as forbidding a landscape for humans as you will find anywhere on the planet. But in 2019 it flooded. It received the biggest douche of water since 1972, and this once-in-half-a-century event was the result of two downpours – a colossal coincidence and for the life of the desert, a glorious one. The previous year had been very dry indeed, but January and February of 2019 brought 11 days of rain. Then in March the system caught the tail end of a tropical storm by the name of Trevor, which gave another four days of serious rain. These events had peak measurements of 448 millimetres and 211 millimetres.

And though it all fell more or less at once, it took its time to travel through the systems of watercourses. The water was a journey – an event that lasted for months – and it sparked a dramatic detonation of plants. That

OPPOSITE
Desert water: satellite view of muddy water flowing into Australia's
Lake Eyre. It only fully fills a couple of times a century.

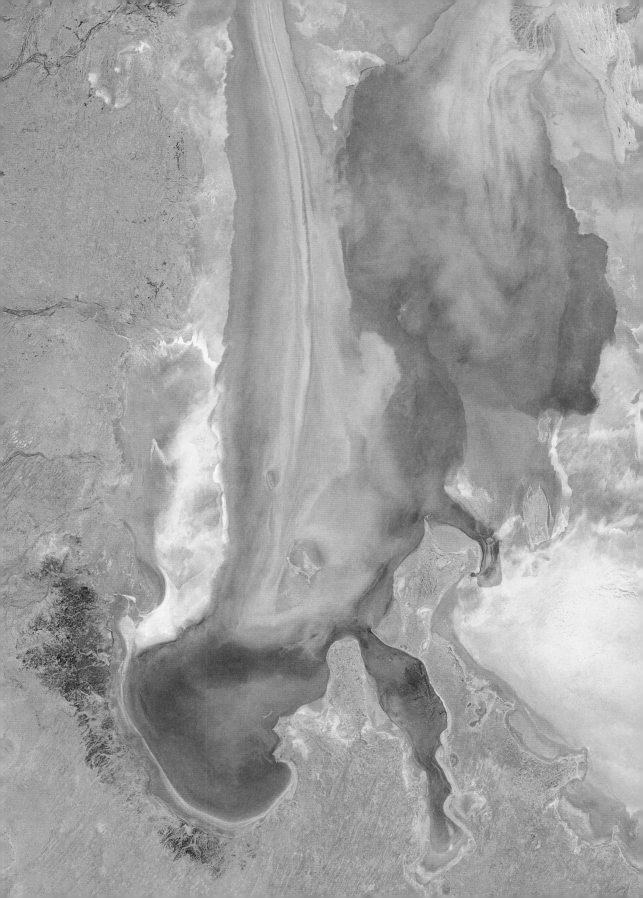

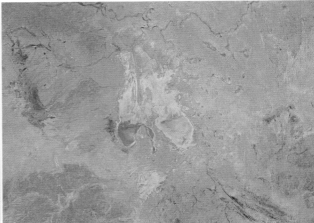

created opportunities for everything else: the desert filled with water-birds, aquatic insects, fish and frogs, all made possible by the largesse of the rain and the plants' lightning-swift response to the changed conditions.

The keystone species here is the coolabah, a species of eucalyptus, the one the jolly swagman sat beneath. It is a welcome tree in the desert, an incongruous one, and its extensive shade and its biological processes, redistributing water and nutrients, give other forms of life the chance to survive. It thrives well on the boom-and-bust sequence of rains, extreme, uncommon and unpredictable. Where rainforest trees thrive on stability, the coolabah has the exact opposite capacity. Not only can it survive the waterless years, but it can also cope with the sudden glut.

One of the problems we have about understanding deserts is our fixed idea of what constitutes desert: sand dunes, camels and the odd pyramid, or perhaps cacti and the bleached skull of a cow. But deserts are various, and not necessarily sandy or even hot. There's also a question of adjustment involved. You arrive in a desert and are shocked by the scanty and parched vegetation. It seems that the whole place is lifeless. But later, after you have seen the extreme parts of your desert, you no longer see the scanty vegetation as deprivation but as a glorious extrusion of life into a thrillingly challenging environment.

ABOVE
Water in the Australian desert: floodwater works its way through channels and fills hollows, eventually reaching Lake Eyre.

OPPOSITE
Keystone species: coolabah trees are essential to the ecosystem of Ethabuka reserve, Queensland, Australia.

OVERLEAF
Destructive and life-bringing: a storm over the Karoo Desert in South Africa.

You can divide deserts into five broad types: the sub-tropical deserts which are, if you like, the default desert, at least in western minds, but there are also coastal, rain shadow, interior and polar deserts. Everything that lives there has the same problem of coping with less water, or less regular access to water than the living things that exploit other environments. Different places bring different aspects and different solutions to the same problem.

We call deserts lifeless because that's how they seem. But we need to adjust our minds to a different concept of the way life is lived. In some desert places you can take up a generous double-handful of sand and imagine that what you have is inert, totally inorganic, and that not a grain of it has anything to do with life. It's hard to believe you have two fistfuls of life: that what you have in your hands is seeds as well as plenty of minerals. You have been burrowing into the great waiting room of desert existence.

We are all used to the idea that plants can over-winter in the form of seeds. It seems only natural. But that's nothing. When it comes to serious waiting, turn to the plants in the desert. Many of them are capable of spending most of their existence in waiting: standing in the wings for year after year, waiting for their cue to act. And the cue of course is water. The cue is rain. It will come all right, but it's best not to be in any itching hurry for it. And so the seeds wait, and they wait for years, with life inside them as a kind of well-kept secret, almost indistinguishable from the grains of dry soil and sand.

And then it rains, and for a brief while the desert is a garden of luxuriance. The rain triggers the seeds, and at once they start to grow. If conditions are kind enough, they will sprout and produce flowers and fruit and the seeds from these fruit will return to the desert when the brief gift of water is taken away again – when once again there is no other option but to wait.

The brief times between dry and dry, between the long expanses of waiting time, can produce in some deserts something that looks as much like a miracle as anything that happens on the plant: not just the occasional plant, doing its best to cope with near impossible conditions, but plants that fill the world from one horizon to the next. One moment conditions were impossible for life: now they are perfect. And in the brief window of perfection, the desert blooms as if it had been the subject of some mad municipal planting project.

It is called a superbloom and, while it is going on, the desert looks like the most fertile place on Earth. So it is – just not for very long. There was a

OPPOSITE
Life-giving violence: saguaro cactus and a storm over the
Sonoran Desert, Arizona, USA.

superbloom in the deserts of California in 2017, and then, unprecedentedly, another followed just two years later, allowing the *Green Planet* film crews to take some of the most extraordinary plant sequences ever filmed. They had an advantage because they knew it was coming: rain in late autumn and early winter had teed it up, and a series of cool days and cold nights made the event a certainty. The desert was getting ready to bloom again: and when the moment was right the place exploded into a series of colour so bright you might have thought they had borrowed them from the palette of Vincent van Gogh.

The spectacle reached a series of peaks, and lasted from late winter to May: a few weeks of impossible glory. The superbloom became a tourist attraction: drive up and see a real miracle. Californian poppies and lupines dominated, along with plants with more fanciful names: owl's clover, California jewelflower, sticky monkey-flower, shooting stars, fairy lanterns and baby blue eyes. Each one used those few short weeks and those scanty and rare resources of water to strut and fret its hour upon the stage, before 'going over', as gardeners say. We use the term 'going to seed' in casual conversation and it implies some sad but self-inflicted disaster. But out in the wild world going to seed is about the life of the next generation. A plant that goes to seed is seeking immortality, its genes living on in its offspring. And so the plants go over and their seeds join the desert, becoming indistinguishable from the land, and so they take up the age-old game of these desert plants. Waiting for rain.

The trick in such circumstances is for the plant to put plenty of seeds out there, and to spread them widely. When seeds are dispersed by wind there is a huge reliance on chance: on factors beyond the plant's ability to control them. It follows that many desert plants will produce plenty of seeds, often tiny and lightweight, and allow the wind to do its stuff. The point here is that, if you have enough seeds, your long-shot chance at becoming an ancestor starts to get closer to a certainty. What's important is getting those seeds distributed in as many places as possible, in the hope that, in at least one place, the circumstances will be right for a sudden dramatic desert superbloom.

The chance of widespread distribution of seeds is dramatically increased by the nature of desert weather systems. Of these the most dramatic – perhaps the most dramatic weather event in the entire world – is the haboob. This is a term borrowed from Arabic and used for a phenomenon found in hot deserts across the world. The most noticeable part of a haboob is wind: a short but ferocious

OPPOSITE
TOP *The desert in flower: yellow California goldfields and orange California poppies at Antelope Butte, Mojave Desert, California, USA.*

MIDDLE *California poppies take the starring role among the wildflowers of the Sonoran desert, Arizona, USA.*

BOTTOM *A little protection: wildflowers around a creosote bush, Death Valley, California, USA.*

OVERLEAF
Unlikely luxuriance: sand verbena and other wildflowers in El Pinacate and Gran Desierto de Altar Biosphere Reserve, Mexico.

storm of moving air. In a dry place where the soil isn't fixed by moisture, the wind picks up sand and dust and moves them across the land in towering and dramatic shapes. Haboobs are notoriously difficult to predict, and they look like the wrath of God when they take place. The plants, of course, have no means of predicting them, but they certainly benefit from the fact that every now and then haboobs happen. And, though a haboob looks like an engine of pure destruction, it is important to remember that there are no disasters in nature, only opportunities. Just as the fall of a rainforest giant is an opportunity for other plants, so a haboob, a traumatic happening for much of desert life, is an opportunity for seed dispersal on an almost unimaginable scale.

The haboob is more than a fierce wind that blows the sand about. It is a complex event that lifts startling quantities of solid material from the surface of the desert and builds from it a wall that can easily stand a kilometre in height, sometimes two or even three. This can travel at a speed of more than 60 mph, and do so on a front tens of kilometres across; 100 kilometre-wide haboobs have often been recorded and there are some that reach a still greater extent.

OPPOSITE
Life spreader: the powerful desert storms distribute seeds over enormous distances; this one in Arizona, USA.

They occur when a thunderstorm takes place above very dry air over
country where there is little vegetation and very little moisture to bind the soil
together. What matters about the storm is the cold it brings with it. The naive
traveller in a warm climate will often think that rain is of no great matter; you
just get wet and carry on. You don't bargain for the sudden and drastic cold
that comes with the rain, not just from the wetness but from the drastic fall in
temperature. You didn't think it was possible to be so cold in the tropics; I
have had a prolonged shivering fit in Africa.

It is this effect, magnified many times over, that creates a haboob. The
thunderstorm produces rain that may never reach the ground, evaporating
before it does so, but while this is going on, a great area of cold air is created
above the desert. This then drops sharply, as it must, creating space above it
for the warmer air to move in. At ground level there is now a wide pool of cold
air that spreads out: a convective cold pool. Along the leading edges of the
cold pool the air is disturbed by the difference in temperature between the
warm air on the ground and the chilled air that has just arrived. This creates

the strong winds that lift up the loose sediment and transport it up into the air: and that is what creates the effect of the great advancing wall. A haboob can be seen from space; similar effects have been noticed on Mars and on Titan, one of the moons of Saturn.

There was a famous haboob in Phoenix, in the United States, in 2011. It was predicted 24 hours in advance, and so was a great event in the storm-chasing community, which is composed of the curious, the adventurous, the scientific, the photographers and the news-gatherers. For most members of the animal kingdom, a haboob is something to get through, to hunker down and survive and hope it goes away soon. It generally does: three hours is about average, though the devastation within that period can be considerable.

But for the plants – at least those that exist in the form of a seed – a haboob is an opportunity of the most majestic kind. We look at the approaching haboob as a wall of sand and dust: but it is also a wall of seeds. The great inconvenience to animals is actually a great bonus to the plants. A seed caught up in such a disturbance can travel huge distances, and those that win the lottery will touch down in a promising patch of desert countryside and once more get on with the process of waiting. One day the rain will come, and for a few brief weeks the desert will bloom, fade and die away again... leaving more seeds, seeds that will get on with the essential aspect of desert life: waiting. Waiting for rain; waiting for water.

There is a way round this, of course. A plant can gather water when it rains, and then store it for later use. Water is precious in a desert, and to have access to it every day is a supreme advantage. But it's an advantage that makes the plant a target: water is precious to everything that lives in the desert and so a plant full of water is attractive to anything that can reach it. It follows that a plant that stores water in the desert needs an impenetrable defence mechanism if it is to survive. And that is exactly what cacti have developed across the millennia.

In a classic piece of Attenborough footage, *Green Planet* filmed David Attenborough demonstrating the efficacy of a cholla cactus's defences. The plant looks so soft and cuddly – at least from a distance – that it has been nicknamed the teddy-bear cactus, but the cuddliness is an unintended illusion. What looks like soft fur is in fact sharp spines. Attenborough put on a heavy-duty leather gauntlet for protection before plunging his hands into the depths of the cholla's defences. This stunt was intended to show viewers the impressive

OPPOSITE
*Not so cuddly: teddy bear cholla plants at Anza Borego,
California, USA.*

number of these weapons; what it actually did was to show how well they work when it comes to causing pain, and how effective they are as a deterrent. It was instantly apparent that Attenborough, despite his gauntlet, was in really quite considerable discomfort. The spines had penetrated the leather without any trouble whatsoever, leaving the gauntlet-wearer gasping with pain. Of course, this being Attenborough, he carried on with the sequence. I'd guess there were three things driving this bravura performance: first, the old pro's certainty that this would be a highly effective piece of television; second, the born teacher's awareness that his discomfort gave his lesson a vividness his audience would not lightly forget; and third, the human being realising that if he stopped now he would have to find a way of doing it all over again.

So we were able to understand, with immense precision, just how good a cactus's defences are, and by extension, just how important that water supply is. Growing those spines is a huge investment of time and energy: it would only be worth doing so if it were a matter of life and death. In a grimace of pain, we saw how the cactus defends its life – very well indeed.

We tend to think that cacti are ubiquitous in hot deserts but, with the exception of a single species, they are indigenous to the Americas, down south to Patagonia, and north as far as Western Canada. The sole exception, *Rhipsalis baccifera,* is found in parts of Africa and in Sri Lanka. There are plenty of cacti

in the Americas: more than 1,700 known species, mostly in places where drought is part of the routine of life and most of them, of course in deserts – they thrive even in the Atacama Desert in Chile, which is the driest place on Earth apart from the poles.

Cacti are succulents, which means that they store water in thick and fleshy parts of themselves. Many plants do that; it's an obvious and economical idea. But unlike most succulents, cacti only store water in their stems. The stems dominate the nature of the plant, so that they seem to be all stem. Almost all cacti grow spines to protect them: botanists use the word thorn for modified branches, while spines, like those on cacti, are modified leaves. These spines protect the plant against herbivores, who would otherwise find their flesh both nutritious and thirst-quenching. They have a secondary use in helping to prevent evaporation: the spines reduce the flow of drying air close to the stem, and also provide a little shade.

The spines may be modified leaves, but unlike normal leaves, they can't do photosynthesis. Though there are some species of cacti that have leaves as we normally understand the term, most don't. Instead, the process of

ABOVE
Is this wise? David Attenborough about to put his hand inside a cholla plant.

photosynthesis takes place in the stem itself. Cacti come in different shapes: there are tree-like cacti with branches, some that form columns and some that live close to the ground and form an almost spherical shape. This contradicts one of the aspects of plants discussed in the previous chapter: the idea that a plant (unlike an animal) goes for maximal outside. You can't go for a more minimal outside than a sphere. By having as little surface area as possible, the plant very effectively reduces surface evaporation. It can't photosynthesise at any great rate, as a tree with a vast surface area of leaves does, but then it doesn't need to. Cacti slow down the processes of life, stay dormant for long periods, live on their reserves – and wait for rain.

As we saw in the previous chapter, the process of photosynthesis involves transpiration: passing water out while taking carbon dioxide in, in a process powered by the light of the sun. Cacti have evolved the ability to take in carbon dioxide during the day – and to store it. This allows them to perform one of nature's great contradictions: they can photosynthesise in the dark. They can eat light when there isn't any. That way they transpire at night, so the water loss is much smaller than it would in the brutal heat of the day.

ABOVE
David Attenborough shows how the barbed spines of a cholla plant can cling to a passing animal and so propagate itself.

They carry out the business of photosynthesis with a delay mechanism that allows them to do so in the cool and dark of night.

Cacti have specialised structures known as areoles, which are modified branches, though they look nothing like what we normally think of as a branch. These produce the spines and, in season, the flowers, which arrive at notoriously long intervals. They are often very showy, in order to attract pollinating animals: insects, especially bees, birds, especially hummingbirds, and bats. Many cacti operate on a very short growing period and a sustained period of dormancy but, during all the times when they are apparently inert, they are in a state of hair-trigger readiness for rain. When rain comes, it is an opportunity not to be missed, and so cacti tend to have a wide, shallow root system that can take in moisture as soon as it is apparent, giving it as little opportunity as possible to sink into the soil or to evaporate. It's been calculated that a big cactus can take in getting on for 800 litres in the course of a single rainstorm. In a way it had better: the water from that one rainstorm may have to last the plant for months or even years. There's something a little like the heartland Olympic sports about this: an Olympic athlete must spend weeks and months getting ready for the one day in four years when it actually matters. The same is true of cacti and many other desert plants, except that the intervals are often irregular, unpredictable and can be a good deal longer than four years. The plants that are ready when the great day comes will win the gold medal: they will survive and become ancestors.

There is a great romance about unlikely places to live, places where the living is anything but easy. You wonder, naively, why any living organism would choose to live in such a desperately inhospitable spot, when there are so many other easier places across the world to live. This is a wrong-headed way to look at things of course; you might just as well wonder why a poor family in a South American earthquake zone doesn't relocate to the Cotswolds. The point is a contradiction – the unlikeliness of life, combined with the inevitability of life. A desert is the classic example of this contradiction: life is highly unlikely here, but also more or less certain. Wherever there is any kind of possibility of life, life will, over the course of Deep Time, find a way of establishing and sustaining itself, no matter how difficult, no matter how ridiculous the proposition may look to our human eyes.

So you get water-filled cacti in a place where there is no water, defending themselves with a determination that made the eyes of David Attenborough

OPPOSITE
TOP *The richness of desert life: golden barrel cactus in California, USA.*

MIDDLE LEFT *A blue-throated hummingbird feeding on the flowers of claret cup cactus, Big Bend National Park, Texas, USA.*

MIDDLE RIGHT *A lesser long-nosed bat approaches a flower of organ pipe cactus, Arizona, USA.*

BOTTOM *Understated beauty: Saguaro cactus at night, El Pinacate and Gran Desierto de Altar Biosphere Reserve, Mexico.*

water, and at the same time you get small mammals that can make a living with the help of one of the most ferocious plants that ever evolved. And if that is triumph for the little creature involved, it is also highly useful to the plant. In a land where living at all is remarkable enough, the teddy-bear and the packrat have found a way of living together.

The packrat or white-throated wood rat is found in many places in the southwest of North America and it is perfectly at home in desert. Its preferred food is prickly pear cactus, whose fruit are also favoured by humans. The packrats also manage to eat parts of the teddy-bear cholla cactus. They are rodents and they make a good meal or a light snack for many desert animals: weasels, bobcats, coyotes, owls and rattlesnakes. That makes them important to the ecology of these dry places; what is sometimes referred to, a little callously, as a larder species. It's fair to say that the packrats don't see it quite that way but, as a species that makes its nest on open ground in land where it is surrounded by enemies, survival is never going to be a straightforward matter.

So it recruits the cholla. The cholla drops small buds, and these grow and allow the cactus to reproduce – to clone itself. The packrat takes these fallen buds and uses them for its own purposes. How it manages to carry them without injuring itself on those gauntlet-piercing spines no one really knows, but the packrat is able to take them up and use them to create a kind of fort. It will

ABOVE
Building a fortress: packrats exploit the cholla buds' protective spines for their own use, but the buds benefit from the dispersal; Sonoran Desert, Arizona, USA.

make a wall of materials – they vary from habitats to habitat, but in country where the cholla grows, the packrats favour them. Often these forts are beneath a bush; if there is a bush available in this arid countryside, it gives added protection against attack from above. Behind the wall of spines, the packrats raise their litters and send new packrats out into the hostile world.

This is a better deal for the cactus than you might expect. The packrats take the buds further than the cactus could ever hope to spread them unaided. Often the packrats carry these buds up slopes, which is beyond the scope of a rolling bud. Some of these then roll away, perhaps aided by the wind, which spreads them still further; others sprout in the walls of the forts. So are the packrats exploiting the cacti for protection, or are the cacti exploiting the cholla for distribution? There is no real answer to that: just the head-wagging marvel of the way that, again and again, life finds solutions to the problems of staying alive. To us, these ways of working together look both impossible and wonderful, but in the course of Deep Time, they are nothing less than inevitable.

When we are children we like to see the world in clear binary terms and so, when we turn to fiction, we like a clear distinction between goodies and baddies. But as we get older, we start to prefer a more nuanced approach, in which even the most hideous of villains has some redeeming feature and the greatest of all heroes is flawed. In sophisticated grown-up novels, relationships between characters are never simple: love is mixed with hate and kindness with destructiveness. Nor is there any break from this when we turn to the wild world. How could there be, since we are a product of it? Relationships are complex and offer no simple answer to the question of who is benefitting and who is being exploited.

That is certainly the case with the wild tobacco plant, which is related to the plant we domesticated for the pleasures of smoking. These plants produce nicotine, which is an effective drug for humans – but the plant produces it to repel plant-eating insects. What humans have used for centuries to poison themselves has been used across countless millennia to poison insects.

The wild tobacco plant produces nicotine when it feels itself under attack, when it knows it is being threatened. These are slightly difficult terms to deal with; we are used to thinking of plants as inert, insensitive things incapable of thought, mere lumps of living matter. But they wouldn't be able to continue

OVERLEAF
Poisonous beauty: flowers of the coyote tobacco plant in Utah, USA.

the business of staying alive without complex and sophisticated responses to the environment in which they have their being. They have to be aware at some kind of level, or they wouldn't be able to function, but the idea that plants have awareness is either one that humans reject totally, as it were instinctively, or one that they embrace too keenly, at a woo-woo level of acceptance. Neither response is helpful, neither is accurate, and both are equally irrational. One way of fobbing off the question is to use a lot of quotation marks: the plant 'realises' it is under attack, so it 'tells' the affected part to manufacture toxins. I'm not sure that really helps: plants lack a central nervous system, sure, but, as we have seen already in many different ways, they respond to stimuli. Their lives would be impossible if they didn't.

It's hard to get an intuitive understanding of the idea; to say ah yes, that's how plants send messages, how one part of a plant talks – or should I say 'talks'? – to another part of the plant. We don't relate to plants in the same way that we do to our fellow animals, after all. But there's no disputing the fact that plants have comms. This was made vivid in a piece of film shot in a laboratory in San Francisco, in which David Attenborough talked to a plant, the plant responded – and we could see the message travelling through the plant. It was as if we had gone at a stroke from the silent movies of plant life into the era of the talkies. In this footage we saw a plant speak or, if you would prefer that in less dramatic terms, we could witness the plant's internal communications system at work and see a

message travel along pathways that have been established for that express purpose.

Attenborough showed us a lab-grown and genetically modified tobacco plant. The genetic tinkering meant that when the plant sent messages around itself, the message pathways fluoresced. So he touched a leaf, in a manner that mimicked the attack on a leaf by a caterpillar – and the message pathways glowed. The message that was sent in the language of chemicals was made visible, so that it was now also a message in light. It was possible to see the messages from the leaf passed around the whole plant: and it was a warning to the rest of the plant: become poisonous. Produce toxins now. Make nicotine or die. And so nicotine was made and the plant survived the notional attack. In the same way, but without fluorescing, the wild tobacco plants in the desert send the same messages when the attack comes. And so they manage to stay alive.

ABOVE
The smell of the coyote tobacco plant attracts a lizard – which then hunts down the plant's main predator, the tobacco hawkmoth caterpillar.

The most frequent predator of the wild tobacco plant is the tobacco hornworm moth and its caterpillars. The moth seeks out the tobacco plants and lays its eggs on their leaves, so that the resulting caterpillars can feast on them when the time comes. That might be considered a serious negative, but here the nuanced nature of the relationship between moth and plant already starts to show itself. The egg-laying moths also drink nectar from the tobacco plants, helping themselves to a substance that gives them power and energy. As they do so, they also pollinate the plants. In other words, the tobacco plant is reliant on its greatest enemy for the continuation of the existence of its own species. The tobacco plants can't have sex without involving their most bitter foe – and at the same time the hornworm moths can't live without the plant that daily tries to poison their offspring.

The caterpillars hatch and begin munching, as caterpillars do, being eating machines. So the plant sends its message around itself and begins to produce the toxins – but that is just the first line of defence. From the bitten edges of the leaf, the plant also releases jasmonic acid, a substance that got is name because it is also found in jasmine plants. That is more useful than it sounds because it attracts big-eyed bugs, who recognise the scent as a summoning signal. It means that there is food available on the tobacco plant, so on they come and they attack the caterpillars, and they also feed on the eggs of moths.

Despite the effectiveness of these defensive methods, some caterpillars survive and grow larger: too big for the bugs to deal with. But the tobacco plant is now able to summon lizards by the same method, and these come in and feed on the now substantial caterpillars. This is a pretty effective system of self-defence. It is complex, highly impressive and evolved over millions of years – but it's still not 100 per cent effective. Some caterpillars will survive, even though they spend their larval lives running the gauntlet of the plant's defensive measures. It's just as well for the plant that its defences are flawed. The tobacco plant needs the adult moths for its continued survival.

Here, then, is a perfect example of the fraught balance in the way the wild world works. One slight tweak one way, and the caterpillars will wipe out the plants on which they depend; one slight adjustment in the other direction, and the plants will lose contact with the animals that are essential to its ambition of becoming an ancestor. The two species are doomed to

operate together: superhero and villain bound together by mutual
dependency, forever in conflict, forever reliant on each other for their very
lives, in a fraught balance and a high complexity that can only be the
products of Deep Time.

It's called coevolution. The idea of coevolution has a kind of magic about
it: the way two (or more) living species affect the way each changes over the
course of the ages. Sometimes this involves relationships of benign
dependency, at other times the relationship is more or less hostile – but it's all
one to the forces of evolution: whatever advantages help their owner to
survive will, if and when passed on, help their offspring too. Sometimes this
relationship takes the form of an arms race: the better your defences become,
the more effective your opponents become in their turn, not as individuals
but over the course of uncountable generations. Sometimes this takes the
form of a greater complexity, but if an increased simplicity is an advantage
then evolution has no false pride about it. We have been schooled to think
that evolution is purposeful, progressive and forever seeking after perfection
– for how else would life have succeeded in producing wonderful us? But the
fact is that evolution works with whatever tools and methods come to hand
– and so when a Chilean mockingbird perches on the top of a cactus and

ABOVE
A battle over spines: Chilean mockingbird at Les Molles, Chile.

raises its voice in song, it sets in train a series of events that represents an arms race between two plants, along with a cunning plan in which a plant has achieved widespread success by doing away with roots, with vegetative stems and with leaves. It has got rid of almost everything by which we define a flowering plant – all, that is, except the flowers themselves. This is the desert mistletoe: you might call it the Cheshire Cat plant. The Cheshire Cat in *Alice's Adventures in Wonderland* vanished until all that was left was the grin; the desert mistletoe has vanished so that all that remains in sight are the flowers. Very nice they look too, to us humans, even if the cacti they have taken on in this arms race might take a different view.

We have seen how cacti grow spines to protect their precious water supply from marauding animals as well as gauntlet-wearing television presenters. The spines also come in handy when it comes to competition with their fellow plants: plants that will happily settle on – and crucially in – a large cactus and take their water supply for their own. The process begins when the mockingbird sings its song. The cactus needs to defend itself from the bird, even though it has no quarrel with the bird itself. It's what the bird brings with its song that causes problems. Like many songbirds, the mockingbird will sing its stuff from a tall, exposed perch. The idea of song is to advertise the singer's claim to the surrounding territory. It makes sense, then, to do this from a good vantage point, from which the bird can be seen and heard, but the bird will prefer a comfortable cactus to perch on, rather than one whose spines are off-putting, so the birds will favour cacti with the least spectacular defences. Birds will often defecate before they take off, since cutting down on weight is always a good thing in air travel of any kind. So the bird will leave its droppings on the cactus and, often enough, encased in its droppings, there will be viable seeds of desert mistletoes. These come from the fruits of the plants, and the mockingbird is a keen consumer of them.

It follows that the best-protected cacti, those with the longest and most forbidding spines, will not be visited by mockingbirds, so they won't have to cope with the mistletoe seeds. The less well-spined plants are the ones that must face that problem. But the seed, once it has germinated, still has the difficult task of reaching the inside of the plant. The longer and more fearsome the spines, the harder that is likely to be. The cacti have developed long spines, and these will often reach 12 cm long so, in order to keep operating, the germinating seeds have acquired the ability to reach out far enough to get into the cactus. Each one puts out a long stalk or probe, technically a haustorium, which stretches out towards the cactus. It grows at a fairly rapid pace, fuelled by the plant kingdom's greatest talent, that for photosynthesis. But once it has got inside the cactus, everything about the invading plant changes.

It becomes a plant inside a plant and delegates almost everything that a plant should do to its reluctant host. It thrives on the water that is stored within the cactus, and also on the energy that the cactus creates for itself by photosynthesis. The mistletoe is a parasite and it needs no more for its survival than a living, thriving, water-rich, photosynthesising host. It operates in the form of thin filaments, a little like the mycelium of fungi discussed in the previous chapter. It gets into the vascular system – the water-shifting system – of the host, and there it lives. It has no leaves, because it doesn't need to photosynthesise; it has no roots, because it gets the water and minerals it needs from the cactus; and it has no vegetative stem, because it gets all the physical support it needs from its host.

It has become so simplified, so greatly reduced a living thing, that all that characterises a plant – all that we think of as a plant – has been refined out of existence. It seems hardly a plant at all, a contradiction of what we think a plant should be. The cactus does the traditional job of a living plant, leaving the mistletoe with few problems other than that of trying to become an ancestor.

Once again there is a human tendency to try and find a moral in this; to ponder on the uselessness of the parasite's existence, or to look for some kind of parable of comeuppance for this presumptuous plant. But of course there is no moral to be drawn, other than the fact that life lives to make life and, wherever life is possible, life will usually find a way of making more life. You can cheer for the put-upon cactus, or you can cheer for the underdog mistletoe: it makes no difference. The mystery lies in the way that such a way of life evolved – and how it manages to continue.

There comes a time in the life of the developing mistletoe that it needs to make more of its kind. It can't do this from inside the fastness of a cactus, however cushy a billet this might be. The plant needs to make more plants, so it needs to have sex. And as we have already seen, for a flowering plant – for

an angiosperm – that means flowers. So the plant that has no roots, or stem or leaves produces flowers. These emerge through the areoles of a cactus, or through any handy tears and breaks in the skin. They produce a mad carnival bonnet of scarlet flowers: something that looks like an outbreak of wild frivolity in the serious and sober-sided cactus. The keep-clear-of-me cactus with its off-putting spines is now sporting a great come-hither signal. In come the hummingbirds. Flowers are not over-common in deserts, and the hummingbirds that are able to make their living in these forbidding places will not turn down an opportunity like this. So they drink their fill from one blooming desert mistletoe, and then pass on to another, carrying the precious pollen as they do so. The mistletoe has an extended flowering period, to make up for the scarcity of pollinators; that again is expensive on the resources of the host. The flowers will then produce the fruits, and then it is

ABOVE
A cheat that prospers: desert mistletoe has taken over a palo verde tree in Arizona, USA.

the turn of the mockingbirds to help both themselves and the mistletoe, by consuming the fruits and spreading the seeds.

The parasite can only exist by performing another of these delicate balancing acts. The mistletoe needs to take a good deal of water and energy from its host cactus to produce the flowers, and when it does so it needs to transpire – to pass out water. That is also very expensive of the resources of the cactus. If the mistletoe takes too much from the cactus, the cactus will die and that, of course, is also fatal for the mistletoe. Killing the host is counterproductive to all parasites; the parasite has a vested interest in the host's survival. So if the mistletoe senses – from the depletion of the resources available to it – that the host cactus is coming under real stress, it will rein itself in, and retreat. It kills off its own external growth and, with its whole self back inside the cactus, it allows the cactus to recover. The mistletoe might have a more successful flowering the next time around; it's better waiting than dying.

The parasitic lifestyle operates effectively in many plants. We have already encountered the jungle monster flower, Rafflesia, which, like the mistletoe, produces only flowers and doesn't trouble itself with photosynthesis. These plants are not closely related; they are an example of similar traits and lifestyles evolving by very different routes: what's called convergent evolution, or just a convergence. There are 3,000 species of parasitic plants in 16 families; one per cent of all flowering plants are parasites.

The rest must make their own way, and in the desert that means the hard way. Any help they can get is gratefully appreciated, even if gratitude is unlikely to be an aspect of the way plants live. We mostly think of deserts as places of unbearable heat, but even cactus-thronged deserts can go through times of unbearable cold – or, in both cases, almost unbearable. In the deserts of Saguaro National Park in Arizona in the United States, it will sometimes snow. It seems almost ludicrously unfair that a cactus, so wonderfully adapted to cope with extremes of drought and heat, should be threatened by water, when it falls in the form of snow, and by freezing conditions. These rare desert snowfalls kill young cacti. But some of these delicate young plants have an advantage: they are protected from the worst of the weather by a neighbouring tree. The cacti that grow beneath a palo verde or a mesquite tree have a much greater chance of survival. That's not parasitism, because

OPPOSITE
The classic desert plant of our imaginations: saguaro cactus in the Sonoran Desert, Arizona, USA.

OVERLEAF
Saguaro cacti can grow 10 metres tall and hold 4,000 litres of water; Sonoran Desert, Arizona, USA.

the cactus takes nothing from the tree; it takes advantage of the tree without compromising the tree's survival or well-being in any way whatsoever. This is a strange and intimate relationship, and there is a good deal more to it, and we'll move on to that shortly.

In the meantime, let us consider the saguaro cactus. You might call this the archetypal cactus – the towering columns with the branches or arms that stick out so dramatically. They are cacti you expect to see in a western: a cowboy riding past one of these many-armed giants pursuing or escaping from bad men, the cactus-filled landscape telling us all how desperately hard things are for all concerned.

But let's look at things from the point of view of the cactus rather than the cowboy. A saguaro cactus can reach more than ten metres in height and, with good luck, live a couple of hundred years. It is, if you like, a living water barrel, and it will expand and contract with the seasons or, rather, with the changing availability of water that comes with the passing of the seasons. A saguaro cactus is by mass a good 90 per cent water, and a big one can hold up to 4,000 litres. This is an extraordinary achievement, or, if you prefer, an extraordinary opportunity. Among those that look to exploit the cacti are – rather surprisingly given our conventional notions of desert – two species of woodpecker, the Gila woodpecker and the northern flicker.

Both species like to nest ten metres from the ground, to make the nest hard for ground-based predators to reach. That makes the saguaro cactus a good bet, for it's hard for the predators to climb those spiny columns. The woodpecker needs to be clever enough and agile enough to reach the surface of the cactus without getting damaged by the spines. This they manage with some confidence, and here they excavate a nesting cavity half a metre deep. This is a deep wound for the cactus to suffer, and potentially fatal. The well-protected inside of the cactus is no longer safe: it is open and the precious moisture inside starts to evaporate. If that goes on, the plant will be unable to survive. But the plant has a strategy for defence: it secretes a thick sap.

This dries out, so the damaged cactus now has a cavity with hard walls. It is so tough that it is called a cactus boot – and it makes a perfect nesting box for a woodpecker, one that no longer threatens the plant. The way the cactus accommodates the bird is good news for woodpeckers, but the relationship is not just a one-way street. The cactus gets something from it too.

The saguaro cactus that stays fit and well will produce fruit: around 150 in a single year. Each fruit will contain a couple of hundred seeds; a long-lived saguaro cactus might produce 30 million seeds in the course of a lifetime. That's a lot of seeds, but a lot of things can go wrong. Perhaps one in a thousand seeds will germinate. The seeds and young plants are threatened by freezing, by drought, by flash floods and by predation of fruit and seeds. But that predation is also good news for the plant.

Birds will eat the fruit and carry the seeds away inside them. That includes the Gila woodpecker and the northern flicker; though woodpeckers mostly eat insects, both these species will take fruit and seeds as well. And when they transfer their activities to a tree – say a palo verde or a mesquite tree – in the traditional manner of woodpeckers, they will defecate and leave nicely fertilised seeds in their droppings. And those that fall beneath a tree have the advantage given to them by the tree. We have already seen that the

OPPOSITE
A safe home: but the saguaro cactus often profits from the activity of the Gila woodpecker; Sonoran Desert, Arizona, USA.

ABOVE
Important resource: A saguaro cactus provides a home for a great horned owl with chicks and a potential home for the excavating golden flicker; Sonoran Desert, Arizona, USA.

tree protects them from snow; it can also protect the plant from direct sunlight and heat. At the same time, the roots of the cactus are able to access some of the moisture that the tree takes from the soil with its much deeper roots. The roots sweat a little moisture into the surrounding soil, and this small loss helps the young cactus to grow and reach maturity.

What we have, then, is another example of the nurse tree: the tree whose presence is a positive asset to their tender young neighbours. We saw this with the balsa tree in the Central and South American rainforest; here, in a quite different environment, we find the same thing in action. A saguaro cactus that reaches maturity will outlive its nurse, and so the cycle continues: another cactus, another bird, another splat of dropping beneath another tree.

So once again we have to adjust our understanding of plants. A plant is not an object; it is a process. It is not thing; it is a narrative, a character in a story that involves many others and is part of many overlapping stories. We have been conditioned to look at plants as things because they don't operate like us humans, like us animals. But that doesn't mean that they don't operate, interact, respond, change. A small weed in your garden is part of a tale that takes a few months in the telling; an oak tree, a rainforest giant or a big cactus tells us a story over the course of centuries. All these stories are about changes brought about through time, and about complex involvement of other living things.

We think of deserts as places where there is very little life, and what there is of it is both remarkable and tenacious. It is characterised by its sparseness. You find a cactus, you don't expect to see another dozen cacti right next door. Deserts are deserted because there's very little water, but also because there are not many nutrients in the poor soil. The fuel for life is there all right, but in very small quantities. As we have seen, deserts are not empty, but the life they support tends to be spread out. Deserts don't teem. Desert life is never dense. Desert plants are lonely. There is no other option.

But the island of San Pedro Mártir confounds your expectations. It stands in the Sea of Cortez, the inland side of the great swooping peninsular of Baja California in Mexico. It is a desert island in both senses: it is deserted, at least by humans, and it is very dry. But here you can find cacti in dense stands, like a field of corn or a squat forest. The lonely plants cluster together as if this were a party, almost within touching distance of one another, in a way unknown on the mainland just a few miles away.

The main plant of the island is cardon cactus, sometimes called the Mexican giant or elephant cactus. On the mainland you can see how they got those funky names: the world's tallest living cactus is a cardon, standing at 19.2 metres. They are related to the many-armed saguaro, but without the dramatic shape. They branch, like the saguaro, but at the base, which can be as much as a metre in diameter.

The island plants never attain such dimensions. San Pedro Mártir is basically a chunk of rock, and it's as remote as you can get in what is almost a landlocked sea. The island is just 2 kilometres long, 1.5 kilometres at its widest point and a bit less than 3 square kilometres in area. So pretty basic and, as you would expect, pretty windswept. Extreme height would not be possible here; a sea gale is no time to stand up and be counted. So the cacti of San Pedro Mártir are somewhat stunted, but they come in eye-baffling numbers.

The cacti can thrive in this way because they have found a way of exploiting the riches of ocean to fuel their lives on land. They can access the protein of oceanic life; not the strategy you would expect from a cactus. They do this by means of their relationship with seabirds, in particular with the blue-footed boobies, which are closely related to gannets and live in very much the same way: hunting for fish in the sea, usually by diving in from great heights. Boobies are birds of the open ocean but, since you can't lay eggs on the water, the boobies, like all other seabirds, must find a hard-standing when they need to breed. Small islands with no mainland predators are what they need, and they come to such places in huge numbers. That is partly because the right sort of islands are pretty scarce, but it's also because many seabirds species choose to be together. They live most of their lives out at sea, at best in small numbers, but at breeding times they become city-dwellers and live in huge gatherings full of movement and noise.

And smell. People who visit a seabird colony for the first time tend to be shocked by the smell. Birds defecate, and seabird droppings stink of fish. Over the years, the stuff builds up. It's known as guano, and in the nineteenth century its virtues as a fertiliser of crops became known in Europe and in Northern America. There was a frenzy for obtaining the stuff, mining it on these offshore islands, often with labour from people living in conditions close to slavery. Guano sparked wars and prompted the United States to become a

colonial power. It was the beginning of intensive agriculture. The fields fertilised by guano could now feed many more people, and that was an important factor in the increase of the world's human population. It's no exaggeration to say that guano – bird droppings – changed the course of history.

Scientists then discovered how to make fertiliser without the help of the birds: we now fertilise our crops with synthetic guano. The world's guano islands were given back to the seabirds, and the droppings started to build up again. On San Pedro Mártir, the process was speeded up when the black rats, which had got onto to the island by way of the guano ships, were eradicated from the island; the rats used to feed on the eggs and the chicks of the nesting seabirds. The nitrogen-rich droppings of the blue-footed booby have fertilised

ABOVE
Nurturing cacti: the droppings of blue-footed boobies nourish the cacti on San Pedro Mártir Island, Gulf of California, Mexico.

these hunkered-down forests of cardon cacti. Here is desert flora fed by the sea: cactus fed on fish, thriving on the life of the ocean.

It is a perfect demonstration of the way one ecosystem impacts on another, the way that an apparently closed ecosystem can nourish another. The classic example is in the salmon run: the salmon leave the ocean where they have lived most of their lives and swim madly upstream to the spawning grounds. In North America, they famously run the gauntlet of bears and bald eagles: all of the fish die and are consumed, but many of them succeed in spawning, and the salmon run will take place again the following year – so the pine forests and the creatures that live in it are nourished by the ocean. Without the ocean's gifts, the ecosystem would be radically different. Out on the island of San Pedro

Mártir, in a less showy but equally extraordinary fashion, the plants form the densest stands of cacti on Earth because, unlike practically every other cactus on the planet, they can harvest the bounty of the ocean.

We have been looking at plants that find advantages from interaction with other species. The way that living things work together to make life possible is perhaps the most fascinating part of the life sciences (or of life itself, if you prefer). But some plants take a different option, and seek to have as little to do with other species as possible, until the time comes for pollination. There are plants that come close to succeeding in their aim of keeping themselves to themselves. Lithops do this by pretending to be a stone.

Lithops are native to Southern Africa and live in deserts, but you would never notice them. They look more like stones than plants. Even a crash-hot field botanist finds it hard to pick out a lithops plant from the real stones. They have become popular as houseplants for their singular appearance and their

ABOVE
Living stone plants: more correctly lithops, disguise themselves to escape from predators. This is a cultivated specimen; they grow wild in South Africa.

OPPOSITE
Stones in flower: a lithops plant, native to South Africa, blows its cover by briefly bursting into flower.

low maintenance, and have acquired nicknames like pebble plants or living stones. They are succulents, like cactus; that is to say they hold their supply of water. That makes them desirable plants for any desert herbivore, providing nutrition and moisture at a single stop. Cacti, as we have seen, protect themselves with spines. Lithops protects themselves by the most elaborate subterfuge.

They avoid being eaten because they avoid being noticed. They hide in plain sight, indistinguishable from the stones that surround them. Neither tortoise nor botanist can pick them out from the real pebbles. They vary in colour, fitting in with the pebbles around them in shades of white, grey, red, brown, occasionally a hint of green, with spots and lines that resemble discolorations found in real pebbles.

You would be entitled to ask how they manage to survive. We have already spent a great deal of time talking about photosynthesis, and the subject will come up again and again as you read on. Photosynthesis operates through chlorophyll, as we have seen, and chlorophyll is green. Pebbles are not often green, and nor are lithops. They aren't parasites, like the desert mistletoes and

Rafflesia: they can only survive by making their own food from sunlight. How can they do this, since their way of life obliges them to avoid greenness?

It is the most elegant contrivance. The pebble-coloured surface of the plant is not opaque, though it looks as if it is. Like the blackened windows of a celebrity's car, the surface of a lithops plant allows light to pass through. Below this window lies the tissue that stores the plant's water; this too is translucent. It allows the light to pass through, deeper into the plant, where the photosynthesising tissues lie. For much of the year this is not a matter of urgency, for the plant spends a great deal of time dormant.

But there are times when there is moisture available, through scanty rains, or, in some species, just by dew. The plants look like pairs of pebbles, and in the split between the pair you find the meristem, the growing point. These can at the right time produce flowers and new leaves, giving the gloriously incongruous sight of a flowering pebble: a brief moment of glory as the plant seeks its pollinators, so that it can produce seeds. The plants spread sideways, so a single plant can look like a series of paired pebbles.

This strategy has a special fascination for scientists. It is not the same as cryptic camouflage (crypsis) in which an organism seeks to become part of the background, the way a brown moth is invisible against the bark of a tree. Nor is it the same as Batesian mimicry, in which a harmless species looks like a dangerous one: the hoverflies in a garden look like wasps, but carry no stings. Lithops are not pretending to be dangerous, and they are not part of the background. They are pretending to be inedible, mimicking a real object rather than melting into a background, but the object itself is not dangerous, simply neutral. There are many similar examples: a stick insect looks like a stick. There are organisms that look like bird droppings and leaves: clearly visible, but not what they seem. It is a strategy of fancy dress, and has been called masquerade. There are about 40 known species of lithops, and there may well be a good few more. It's hard to tell – because they look so much like pebbles. They baffle scientists and predators with equal facility.

The baobab tree is every bit as singular as the lithops plants, but on a rather more massive scale. Their trunks can have a diameter of 14 metres: in deserts they are not just local flora but local landmarks. They are long-lived: a thousand years is routine, twice that far from unusual. Their dramatic nature has created a slew of legends: the most popular is that the trees were so proud they offended the gods, who turned them upside down and forced them to grow root-upwards... and that's exactly how they look in the nine months of the year they are leafless. Beneath the branches there is a great barrel of a trunk, one that can store water in prodigious quantities. This is no secret; human hunter-gatherers have raided baobabs for a drink across the millennia, and elephants still do. The baobab has an effective defence: it can regrow bark. Savage wounds that would kill a normal tree are a mere inconvenience to a baobab. Elephants are often profligate with trees in their quest for nourishment, and a complete ring-barking means that the tree can no long transport liquids around its system and must die. But a baobab can carry on: you often see baobabs with horrific scars, still growing imperturbably, some fresh, others of centuries-old and long-healed.

But this balance has shifted. Baobabs are sustaining more and more damage from elephants, and they are not recovering. A number of ancient and revered baobabs have been killed around Mana Pools National Park in Zimbabwe. The elephant attacks have become more frequent and more violent, and the trees are suffering. That's bad news for the elephants as well as the baobabs: they are destroying their own fallback water supply. The

ABOVE
Challenging times: a desperate elephant attacks a baobab tree for sustenance in Zimbabwe.

reasons for his new behaviour are complex, but they come down to two points. The first is the ever-expanding human population, which requires ever more space. Elephants also need space. In arid country they traditionally make long migrations to take advantage of seasonal availability of water and associated greenery. The wild routes and the wild places are now damaged and destroyed. The second problem is that there is simply less water around. The dry season is longer, the water is scarcer, and so the elephants are forced to turn to their emergency supplies more often than they did before. Now that is beginning to run out. Climate change is not one disaster; it is a series of linked disasters, many of them quite unexpected, taking place all over the world. Even the desert species are suffering.

It's impossible to withhold admiration for the plants of the desert. We can't help but identify with them, anthropomorphise them, see them as tough guys in a tough environment, battling underdogs who make it through because of their inner strength, almost as if this were a test of character rather than an example of evolution. Making stories full of vivid characters is the way we humans seek to understand important matters, and a giant saguaro cactus is certainly a perfect image of the tough-guy plant. If you can live and prosper in the desert, you can cope with just about anything, after all. If there's one group of plants you'd back when it comes to coping with the crisis of global heating, it's cacti, especially the many-armed saguaro. When everything else is parched and drying and dying, surely the monster cacti will cope, and perhaps prosper as never before.

But this is not the case. Climate change is a complex and elusive subject, and it's hard for us to find an easy, intuitive understanding of what it means – perhaps because we can't turn it into an easy story with vivid characters. There are no obvious good guys and bad guys; we are all, to greater or less extent, to blame. Which makes it a rotten story, so far as most of us are concerned.

But we need to find a way of understanding the crisis, for our own sake and for the sake of our great-great-grandchildren. Perhaps the idea of the suffering cactus will help – so let us turn again to Saguaro National Park in Arizona in the United States, which was declared a National Park back in 1933, mostly to protect the landscape and the cacti that made it a thing of wonder. The current state of the park has been monitored across the years by means of repeat photography: if you take a photo from the same spot at regular intervals, you can compare the images and see – easily and vividly – the changes that have taken place.

So in a classic piece of juxtaposition, put together for the filming of the television series of *Green Planet*, David Attenborough stood in a section of the dramatic landscape, one that had been photographed in 1935. It was clear from the contours of the hills that the two images had been taken from precisely the same point. The first, in stark black and white, reveals a fine forest of cacti: not the stunted plants of San Pedro Mártir, but a fine and flourishing regiment of classic desert flora, standing tall above the arid surface – one of the great wonders of the natural world. But alas, in the

OPPOSITE
Under attack: a baobab tree after repeated assaults by elephants; Mana Pools National Park, Zimbabwe.

OVERLEAF
A thousand years is nothing much to a baobab tree.

luxuriance of modern colour reproduction, Attenborough was shown
standing in a stark landscape with very few plants to relieve it. And more
than anything else, it was the contrast between the two images that was
stark.

Cacti populations are not stable in the way that trees in a forest are
stable; the numbers wax and wane, and the saguaro cacti do so over a
timescale that humans can't easily relate to, over the course of a century or
so. The Saguaro National Park is currently in a waning period – but that
means there should be many young plants, below the level of those that
dominate the 1935 image. The desert should now be a great crèche full of

baby cacti, getting ready to take their place in the grown-up world of the cactus forest. But alas, this is not happening.

There are always problems when you look for causation in the natural world. The system is too complicated for the simple answers that please human minds. You'd have thought that the desert would be an exception; that, in this harsh world with little water and very few living things, the issues would be straightforward. As we have already seen, nothing about the desert is simple: even here, the continuation of life depends on many factors involving many different kinds of living things. So we must be aware that some of the problems for the cacti come from the shortage of nurse trees.

ABOVE
Desert survival: David Attenborough among the saguaro cacti in the Sonoran Desert, Arizona, USA.

We have seen that the saguaro cacti do a great deal better when they are protected by one of the sparse desert trees. Before the place became a national park, humans who ventured there routinely damaged and destroyed trees for firewood and other purposes, and the law on unintended consequences had helped to create the desert crisis.

But the fact is that deserts all over the world are getting hotter and drier. That might not sound like a problem for cacti, but it is: they have not evolved to tolerate limitless heat and aridity. The saguaro cacti evolved for the conditions in which they were photographed in 1935. Since then, here as elsewhere, humans have shifted the evolutionary goalposts. Sure, the greater warmth is reducing the occasional killing freezes, which can wipe out a generation of seedlings. But the greater temperatures create an increased demand for water in the hot times of year. The seedlings can't hold as much water as more mature plants. If there are longer and greater droughts along with higher temperatures, the lack of water will cause the numbers of cacti to

ABOVE
A time of plenty: a crowd of cacti pictured in 1960 at Saguaro
National Monument, Arizona, USA.

fall. The aggregate annual temperature in the southwestern United States has risen by 1.2 degrees between 1950 and 2010. It's been said that this is like moving the whole park 150 miles further south. But that is not, of course, the end of the matter: it's been calculated that temperatures will rise by between three and five degrees by 2100.

There are some signs of hope. It's been found that there are smaller declines in the cacti that live on steep rocky soles, where the rainfall is less likely to evaporate at speed. But the situation is deeply troubling. Wherever you find them, cacti are becoming one of the most at-risk groups of plants in the world in the time of changing climate. The saguaro cactus is central to life of the desert; more than 100 other species depend on them for a living. That makes it a keystone species: remove it, and whole arch of life in the same environment will collapse. We have reached a stage when the tough guys need protection.

ABOVE
A time of depletion: David Attenborough in the spot where the previous picture was taken 60 years earlier.

SEASONAL
WORLDS

Those of us who live with the seasons find nothing extraordinary about the way that every year we live through turns and turns again. It seems to be the natural and inevitable rhythm of life: when we look for something remarkable, we turn to deserts and rainforests. The homely transformations from winter to spring, into summer and autumn and back to winter again – these please us at a profound level, but they seldom strike us with astonishment.

Perhaps they should. If we visit a favourite landscape with fresh eyes, and do so from one season to the next, we see changes so drastic, so radical that you might as well be in somewhere completely different. A place of lush warmth and contentment, humming and buzzing with life, might six months later be a white wilderness. A bleak grey landscape can turn green in the course of a few weeks, and take on a completely different set of colour values. It will also sound completely different: a soundscape of whistling wind and the occasional croaks of crows can become one filled with birdsong and the buzz of insect life. A place can change from friendly and welcoming to hostile and intimidating. A place full of the joys of being alive can become a place of death, where everything you see is either dead or dormant. In such a spot, life seems to turn itself on and off like a tap.

To survive in these places of colossal contrasts, you must be able to encompass the seasonal rhythm in all its changing moods. You must find a strategy to deal with extremes. We are accustomed to wonder at the life of the seashore, which must cope with the twice-daily shifts of the tide: everything that lives there must be exposed to the sun and the wind and then drown in the depths, day after day. But the annual shifts of temperate lands are perhaps still more extreme. What can be more unlike a soft summer day in a butterfly-thronged flowering meadow by the edge of a leafy wood – the air filled with the song of birds and the ground marked by the signs of passing mammals – than the same place six months later in a wind that has swept in from the pole, not a leaf to be seen on the trees, no flower on the ground, no insect in sight and the few birds you can see are thinking of nothing but the possibilities of survival for another day?

The miracles that lie under our own noses might in the end seem the most remarkable of all, because appreciating them involves a radical shift of perspective. The tree you pass every day of your life on your way to work or the shops is a miracle to be compared with the towering dipterocarp in the

OPPOSITE
Change follows change: oak leaves in a winter stream;
Minnesota, USA.

Borneo rainforest and the saguaro cactus in American desert. Familiarity breeds indifference rather contempt but, for those of us who live in temperate lands, this chapter will be about the miraculous nature of everyday life. That tree you passed has survived perhaps a hundred winters, gone through a hundred springs to a hundred summers, and every time it has done so it has changed from top to bottom. How can it cope with such changes? How can it deal with the ever-shifting seasons? Such a tree requires a versatility no rainforest tree can match. Almost every plant of the seasonal lands must be a master of change: able to change shape, colour, and way of life. Some survive by sleeping through the bad times, others die... only for their offspring to rise again in spring.

ABOVE
But change will come: snow-locked trees, Kuntivaara, near the Arctic Circle in Finland.

The evergreen plants have a different strategy. When I visited the northern conifer forests in spring, I dressed for an exceptionally cold winter's day in England – and found I was at least one layer short. This is a hard climate, and the trees that survive the winters – hanging tough into the spring, year after year – need special adaptations. Here you find the great conifer forests, sometimes known as the taiga, or the boreal forests. The pointed Christmas-tree shape that we all know so well is an adaptation: it allows the snow to slip off easily without taxing the tree's strength. This doesn't always work. The combination of a warm front bringing wet snow, temperatures just above freezing, and a big wind can deposit huge, potentially damaging quantities of snow on a tree. The leaves of conifers

don't look like leaves, so much so that we call them needles. These possess adaptations that resist water loss, and that's why the tree can save energy by keeping them throughout the winter; more on this a little later. An important advantage to this strategy is that the tree is already fully leaved when spring finally arrives. These northern conifers – trees that bear cones – are able to alter their internal biochemistry to resist freezing. They can cope with a high level of dehydration. They have thick bark to insulate the more delicate systems inside. The pinecones protect the seeds within them. What's more, the forest itself protects the trees: the warmth created by the dense forests allows the trees inside it to make it through the winter.

Winter, then, is mostly a time of waiting. Of not dying. Those that succeed in the monumental task of not dying through the brutal winters – for the further north the more brutal they are – must be ready to go the moment

ABOVE
Season of growth: a maple sapling among wood anemones in April; Neubrandenburg, Germany.

the living gets easier – because it won't be easy for long. You can look on the entire northern winter as an extended 'hold', as athletes say: the time when the great athletes of the Earth are all crouched on their starting blocks, unmoving but ready – waiting for the starting gun of spring – and, as the British sprinter Linford Christie used to say, you need to go on the B of the bang. Those that get left behind are the losers. The further north you go, the narrower the window for reproduction. Long waiting must be followed by rapid response and decisive action. The trees must switch from dogged endurance to the most hectic activity in the year, and do so at a great pace. One moment, nothing, next moment, everything. The snow melts, and it's now or never. Bang!

There is a geographical point at which the optimal strategy for a tree changes. You can mark this with a line on a map, so long as you remember that it represents a simplification of reality. But there is a notional line: north of it you find mostly conifers; south of it, mostly deciduous trees, those that lose their leaves in the winter. It follows, then, that these deciduous trees must be prepared to grow their leaves again as the season changes. They must be ready to go on the B of the bang – while remembering that a false start does them no good at all. If they responded avidly to every freak warm day of winter, they would find trouble the instant the cold returned: nipped in the bud, quite literally, as the frost and the extreme cold killed off the vulnerable growing tips. So, instead of going off half-cocked, the trees wait for the real start of spring, rather than the false promise of unseasonal warmth. No one is quite sure how the mechanism works, but they count down the weeks, responding to the ever-lighter and ever-longer days.

It's worth reflecting here that a deciduous tree changes what it is – changes its essential nature. It becomes a quite different sort of living thing from one part of the year to the next. Animals can change their diets, and crucially many animals, especially birds, can migrate: they can change their place, swap one environment for another. A tree, unable to migrate, must, in the lands of extreme change, swap one way of being for another. (It's worth adding here that if you change the climate the strategy can't help but go a little haywire: like pouring a glass of water over your computer.)

But as the spring arrives, the trees spring into action: for action it is, despite the trees' unmoving nature. They radiate heat that melts the snow,

and they absorb the heat of the sun. As we humans profoundly relish such an experience of 'feeling the sun on your back' after a prolonged cold spell, so the trees respond in a still deeper way. The sun brings life: the trees come back to life after the pretend death of the winter, in the manner of the Sleeping Beauty, with the sun playing the part of the handsome prince. For a tree, winter is retreat. The law of maximal outside is revoked and the trees lose their leaves, and basically stop functioning as living plants, for without leaves they can't photosynthesise. Instead, they store. There is nothing left that is easily accessible to the life of the forest; it is all stored underground in the root system, water and sugars. When the starting gun goes, the tree must transfer its life from the roots to the tips, and this rises in the form of sap – water filled with life-giving sugars. The sap must rise to give the plant the energy it needs to make new leaves. Once the new leaves have been established, the plant can respire and make food once again. It rises under

OPPOSITE
Seasonal resource: a yellow-bellied sapsucker on a sugar maple tree in Texas, USA, as the spring sap rises.

ABOVE
Wells full of energy: a yellow-bellied sapsucker by the holes the birds make in the search for sweet sap.

pressure from the water stored in the tree's roots, which forces the sap upwards. The tree – the whole forest – is stirring back to life, and it's all to do with the rising sap, and that is triggered by the lengthening days and the greater warmth of the sun.

The sap is new life for the tree, and it's also life for anything else that can get hold of the stuff. The trees don't go out of their way to make this easy, but some animals can get through the tree's defences, and that is crucial to the continuing life of the creatures of the forest. In the forests of North America, around the notional line where the deciduous trees take over from the conifers, the sugar maple trees are sought out by yellow-bellied sapsuckers. This is a species of woodpecker but, instead of feeding primarily on insects and other invertebrates in the tree bark like most woodpeckers, these hammer deep into the trees to reach the sap. They create a series of wells neatly drilled into the trunk of a good tree.

The sap is the only food resource of the forest in the brief weeks of the beginning of spring, before the insects, buds, leaves and flowers appear, so it is an important resource for many species. Ruby-throated hummingbirds arrive before there is nectar to drink: they must find enough sap to subsist on or die. They exploit the wells made by the sapsuckers, and will defend them from rivals of the same species with great vigour, for life depends on doing so. Chipmunks, squirrels and bald-faced hornets also exploit these holes, while bears occasionally emerge from hibernation to feed on the sapwood on a more massive scale.

The tree defends itself by sealing off the holes; the sapsuckers counter this by reopening them. The trees awake with the sun and the warmer days, and their new life is made accessible to others by the sapsuckers. The sapsuckers lead the plunder, but the trees can sustain the damage – just about – and still grow and prosper, so that the following year they can be plundered once again, and life can sustain itself from one season to the next and from one year to the next. Such systems are like a master juggler who has all kinds of beautiful and fragile things flying from hand to hand, twisting and tumbling in the air in a way that looks so smooth and effortless that you might think that nothing could ever go wrong. But a tiny miscalculation from the juggler and the whole system collapses, the objects fall to the earth and shatter and the juggler can no longer continue the act.

OPPOSITE

TOP *Other species benefit from the sapsuckers' work, like this squirrel.*

BOTTOM *No flowers yet: but the ruby-throated hummingbird can survive by exploiting the sap-wells.*

As the starting-gun of spring fires, everything that hopes to live through the season and beyond must find or hold a place. A territory. We are used to that concept among animals: in spring, birds sing to establish and safeguard a place where they can feed, mate and rear young. Mammals like otters leave their droppings in prominent places to inform others of their interest in this place. Plants must do a similar sort of thing: claim a place for their own. After all, no place, no plant.

The business of place is even more important to a plant than to an animal: once it is established, a plant can't move. It is stuck with what it's got; it can't try somewhere else down the road that looks a bit better. So it must do all it can to make the most of the spot it finds itself in. We have already seen something of the competition for place in the rainforest, when a giant falls and creates a gap. The arrival of spring is the same kind of opportunity in the seasonal places of the world, but more predictable. So let's look at a patch of open ground in England. Life is put on hold in the cold months, but the plants must strive with each other to make the most of the opportunities the place presents when the warmth returns. Ivy, bramble, hops, nettles and many others race upwards toward the life-giving light, often using each other as support. For them it is a competition in which failure is death; for those of us who watch, it is a glorious overflowing of life. Charles Darwin wrote about just that prospect, in the ringing conclusion to *On the Origin of Species*, celebrating a 'tangled bank' full of life of many different kinds, all of it operating on the principles he had so painstakingly outlined. 'There is grandeur in this view of life,' he proclaimed, and rightly so.

And that view of life includes the cheats. We celebrated the desert mistletoe in the previous chapter; in England, a plant called dodder operates on much the same principle. (It also does so in many other places; there are more than 200 species of dodder worldwide.) You will often find dodder on a tangled bank: a perplexing mass of red strands that seem to have no very good reason for existing. There are some good folk names for the plant: strangle tare, scaldweed, wizard's net, devil's guts, pulldown, witch's hair. The visible strands are attached to the plants around them, and from these plants they obtain the water and nutrients they need. It's clear from the colour – red, not green – that they don't photosynthesise: their leaves have been reduced to tiny scales. They seem to be just threads. But they can produce flowers and many tiny seeds, which are carried off by the wind.

Once a dodder seed has germinated, it must find a plant within five or six days. If it fails, it dies. But it doesn't rely entirely on blind luck. The young plant is capable of locating a potential host plant by picking up chemical clues, and then growing towards it. It's a little like a sense of smell: the plant responds to chemosensory stimuli and acts upon them. Laboratory experiments have shown that the dodder is capable of distinguishing between potential hosts. Given a choice between a tomato plant, a very good host, and wheat, a very poor host, the dodder will grow towards the tomato. When the tomato plant is removed but its volatile chemicals retained, the dodder grows in their direction – sniffing its way to success.

Once attached to a host, the dodder penetrates the stem by means of haustoria, just as the desert mistletoe does. These penetrate the water-

ABOVE
Letting others do the work: the parasitic greater dodder on bell heather; Exmoor, England.

carrying (vascular) system of the hosts, and from then on, so long as the host survives, the dodder prospers. Its roots wither swiftly, for the dodder has no further use for them. The plant twines around or sprawls across its host, searching for neighbouring plants. When it finds them it will attach itself to these as well. In this way a single dodder can exploit a number of plants at the same time, countering the vulnerability of the eggs-in-one-basket strategy. It's also been found that dodder plants don't flower of their own accord. Instead they time their flowering to coincide with that of the host; basically, they are able to eavesdrop on the host-plant in order to exploit it all the more efficiently. But plants affected by the same dodder can use the parasite as a line of communication. They can transmit the information that they are being eaten by caterpillars, an advance warning that gives their neighbours a chance to set up their defences and make themselves distasteful to the caterpillars. The mighty Darwinian principles are as vivid and as active and as all-powerful on an English bank as they are in the tropical forest and the desert.

ABOVE
Parasitic beauty: the greater dodder in flower; Runnymede, Surrey, England.

Time is as important as space. In the lands of seasons, timing is life and death. There is a time when the availability of warmth, light, water and nutrients are all at their peak, and this comes as spring shifts gear, the days get longer and the weather is reliably warm. The problem for each individual plant is that this magic moment is not exactly a secret. It follows that when the right time comes along, many, many plants will try to take advantage of it. So, in a kind of ecstasy of competitiveness, many flowers bloom all at the same time, creating some of the most pleasing spectacles that any human ever set eyes on.

It is as if the world is showing off, doing all it can to charm the eyes of its most dominant species. There are spectacles all over the seasonal world that bring these immense detonations of colour and scent, as if the planet was telling us, 'We aim to please.' In England, people make pilgrimages to bluebell woods, places where the woodlands explode with blue – the bluest blue that anyone has ever seen. People take the boat from the Pembrokeshire coast to the island of Skomer to see the annual spectacle of the bluebells, as if

ABOVE
Successful exploitation: the toothwort avoids the chores of photosynthesis by parasitising an ancient hazel; the flower is the only visible part of the plant; Peak District National Park, Derbyshire, England.

an entire island had turned blue for the simple purpose of giving pleasure. All over the world there are places with wild floral abundance: lupins in New Zealand and Iceland, bluebonnets in Texas, cape snow in South Africa, and apricot blossom in China.

And the point is that that it really is all there to please. There is competition not just for heat, light, water and nutrients, but also for pollinators. The flowers are there to attract the bees and all the other pollinators: the basic job of a flower is to be attractive, its biological functions to be gorgeous. The fact that we humans also find flowers attractive is perhaps not entirely coincidence: flowers are supposed to be attractive, but the roots of plants are not. There is no percentage in gorgeousness for a root, and few people choose to wear a root as a corsage or buttonhole. In flowers we have natural selection for beauty: survival of the loveliest. A flower's beauty is something that many species of the animal kingdom respond to, as

ABOVE
Detonation of beauty: people travel across the country to see the apricots in blossom in Huocheng County in Northwest China.

OPPOSITE
Competing or cooperating? A bit of both: sand bluebonnet and paintbrush flowers; Hull Country, Texas, USA.

OVERLEAF
When the time is right: a bluebell wood at its flowering peak; Belgium.

they seek nectar and pollen. Sometimes – often – this is to the plant's advantage.

The sense of competition underpins these impossible spectacles of beauty, but it is more complex than that. As the plants compete for pollinators, there is also a sense of them working together. A massive number of plants, all in flower at the same time, will, all things being well, attract huge numbers of pollinators. It is in every individual plant's interest that the pollinators should visit plants other than themselves, but they must be plants of the same species, or the sex that makes more plants won't happen. So the great blooming spectacles of the world are both competitive and cooperative: one for all and every plant for itself at the same time.

But how do they know? That remains one of the great questions about plants that have their being in the seasons. They need to have some kind of winter memory, in order to fire what scientists call their 'vernalisation response': that is to say, to come to life in spring. This ability to bide their time and wait for the right moment has been called the cold clock: a response

to low temperature and the length of days. In seasonal lands the varying length of days is as important as the changes in temperature. In the Shetlands, islands off the northern coast of Scotland, summer will bring more than 19 hours of sunlight in 24; in winter fewer than six hours of daylight.

Some seeds won't germinate before they have endured a long period of cold: after all, there is no point in sprouting optimistically on a warm day in November. The seeds have the ability to count down the days and respond to the clues of increasing daylight, so they can rise from earth at the right moment to catch the sun, the warmth and the peak abundance of pollinators.

The plants that reject the haphazard nature of pollination by the wind have created a crisis of dependency. They require the intervention of animals – a great deal of the plant kingdom depends on the kingdom of animals. That is not such a big deal as it sounds: as we have already noted, the entire animal kingdom depends, directly or indirectly, on the kingdom of plants, while the kingdom of fungi is also essential to the continuation of life in all its forms, as we will see later in this chapter.

A plant must do all it can to attract animal pollinators. It must create a flower of appropriate loveliness, perhaps with a scent that broadcasts the message of its loveliness still further, a particularly useful trait if you are looking for night-time pollinators. It helps if you can offer a life-giving draught of nectar. But if the visitors that you attract are there for nectar alone, you must ensure that they still do the job of pollination. If they just drink the nectar and go, the plant has wasted a huge amount of energy without doing itself any good.

As a result of this, some flowers have adopted the strategy of explosive pollination: showering visitors with pollen which, if all goes well, they will transport to another flower of the same species. The mucuna plant performs just such a trick when it is visited by nectar-seeking fruit bats. The plant is found in the old world tropics and is often grown as fodder crop; it has been called monkey tamarind and Yokohama velvet bean. The action of a feeding bat causes the flower to spring open, exposing the anthers and stamens, and therefore the pollen. Surprisingly, the plant continues to thrive north of the tropics into Japan, where there are no fruit bats to do the job. Here the monkeys, Japanese macaques, have acquired the skill of opening the flowers by hand, and this also triggers the release of the pollen. A strategy that

evolved for one kind of animal works just as well in quite different circumstances. You find such happy accidents again and again in evolution.

Some species of bumblebees have evolved such a close relationship with the plants they feed from that they are able to summon them into action. They can trigger the pollen release of these flowers by the nature of their buzz – by the frequency with which they move their wings. They are able to alter the frequency to suit the flower, and it's a trick that humans can replicate with the help of a tuning fork. The phenomenon is called buzz pollination or sonication: the plant responds to the bee's summons and releases pollen. Some of this will provision the larvae back in the hive, but enough will make it from flower to flower and allow the plant to reproduce, a classic example of fair exchange and mutual profit. Honeybees, including the

ABOVE
Mutual dependence: Pallas's long-tongued bat feeding on an ox-eye vine in lowland rainforest, Costa Rica.

bees that have been domesticated by humans, are unable to perform this trick. Insect pollinators are not interchangeable. Honeybees are poor and unreliable pollinators of tomatoes, blueberries and cranberries, for example. The presence of wild pollinators is important: they can't be replaced entirely by bussing in domestic bees, an initiative taken in many parts of the world where wild pollinators have been wiped out or seriously depleted by overuse of insecticides, most notably the United States.

Attracting pollinators is crucial to the continuation of many plants. Few take more elaborate steps to bring them in than the orchids. The members of this group seem perpetually prepared to produce flowers that go almost too far in their elaborate quest for pollination, evolving an intense and often dramatic relationship with the species that spreads their pollen. One of the most extreme examples is found in the kwongan bush area of southwest Australia.

The Mediterranean climate, or Mediterranean-type climate, is a classic form of seasonality. You can find such a climate in many places around the world: cool wet winters and hot dry summers. All forms of seasonality involve extreme contrasts: this is just a slightly different set of extremes than those experienced in England and the northern United States. You can find a Mediterranean climate in the 21 countries around the Mediterranean, as you would expect, and also in California, Baja California in Mexico, Chile and Argentina, South Africa by the Cape, Western Pakistan, and in southwest Australia. Here the kwongan heathlands occupy an area the size of England, and they possess an astonishing diversity of plants: more than 7,000 vascular plant species, compared to around 1,500 in England, of which 80 per cent are found nowhere else, compared to just 47 endemics in England.

In such places you tend to find extreme forms of adaptation, and the hammer orchids effortlessly fit into that category. The soil of the kwongan is poor, so the plants must stockpile nutrients and play a waiting game. Timing is the essence of life in every seasonal habitat, and flowering at precisely the right moment is a life and death matter. The extraordinary grass trees send out their spikes of flower when the conditions are right. This flowering coincides with the emergence of small wasps that seek out the nectar. So far so normal. But the hammer orchid has come up with a more intense strategy to lure those same wasps away from their rightful target.

OPPOSITE
TOP LEFT *Buzz pollination: a Chilean bellflower responds to stimulus and releases pollen.*

TOP RIGHT *A snowdrop releases pollen after stimulation from a tuning-fork.*

BOTTOM *A metallic green sweat bee pollinates a horned nightshade while feeding on nectar: the buzz of the wings releases the pollen; Arizona, USA.*

The flowers imitate both the texture and the scent of the wingless female wasps. So intense is this deception that the males, entranced by the parodic females, prefer the imitation to the real thing. The orchid flower is equipped with a hinge; when the male grabs the wasp mimic and tries to fly away with it, looking to copulate, his forward motion causes the hinge to bend and the wasp is lifted up and over the flower. He is then hammered, upside down, against the single stamen. Two pollen sacs or pollinia stick to the wasp's back. He tries repeatedly to wrest away the supposed female, but is continually smashed against the back of the flower. He lets go eventually and, having recovered from this adventure and desire rekindled, he makes a beeline for another hammer orchid. Once again the wasp finds himself upended and hammered, this time depositing the pollen sacs onto the pistil of the new plant. Beaten up and frustrated, he continues his sexual

ABOVE

Perfect position: a warty hammer orchid has a hinge mechanism that manoeuvres insects into the optimal place to receive pollen; south of Perth, Australia.

adventures – only to confront another orchid. The males get nothing, the females stay unfertilised and hungry. The orchid's brilliant power of mimicry has brought about this annual triumph.

This is bad news for the wasp on a number of levels, not least because the females, being wingless, require the males to give them a ride up to the summit of the flowering grass trees where they can feed and lay eggs. But the long-term survival of the orchids requires the annual emergence of gullible wasps: to fool the whole lot of them and so prevent them doing their stuff with a real female would be counterproductive. This doesn't happen because the orchid's flowering season is short, much shorter than that of the grass trees. Eventually the male wasps catch on. No longer duped by the orchids, they turn – perhaps as second best – to real females. The generation that emerges the following year will be ready to meet the demands of the hammer

ABOVE
Duped into an orgy: male wasps on a hammer orchid, tricked into believing it's a female wasp, do the work of pollination in the Jaarra Forest, Western Australia.

orchids for the brief time of their flowering. Then once again they will turn their attention to real females and the grass trees.

The competition for pollinators is perhaps even more dramatic in the Eastern and Western Cape provinces of South Africa. This is where you can find the fynbos flora. Here, in another Mediterranean-type climate, there is another astonishing detonation of floral diversity. The species are so many and so distinctive that this comparatively small area is considered to be one of only six floristic regions (sometimes confusingly called kingdoms) in the world and the most floristically diverse place on Earth. It is found in a narrow belt, never wider than 200 km, that stretches from coast to coast, occupying about half the land area inside that belt – and containing 80 per cent of the area's plant species. Estimates of the diversity vary, but there are getting on for 9,000 plant species that make up the fynbos flora, and nearly 70 per cent of them are found nowhere else in the world. This includes the rooibos or red bush, which is cultivated to make an excellent caffeine-free tea.

It follows that the demands of timing in this strongly seasonal area will create an annual spectacle of simultaneous flowering, and it is a matter of taste

ABOVE
Cleansing and renewing: fire sweeps through fynbos flora in Cape region, South Africa.

OPPOSITE
A fire lily, Cape region, South Africa.

and local patriotism as to whether or not you think it beats that of the kwongan. The fynbos flowers compete (and perhaps cooperate) to bring in the pollinators and put them to work: as extravagant a display of floral sex as you will find anywhere in the world. But even as we celebrate this bewildering display of biodiversity and bioabundance, let us also take note of an outlier: the fire lily. Can there possibly be a better way of competing than not to compete at all? This is the drop-out flower, the plant that has opted out of the system – but only to exploit the system with as much efficiency as any of the other players in the fynbos system.

The clue is in the name: in the fire. Fynbos is a heathland system: dry summers and poor soil create a flora of low-growing woody plants in an open landscape. It is not hard to see that such a landscape is susceptible to fire. A

ABOVE
Fire as opportunity: a fire lily can wait 20 years below ground before taking advantage of a fire; Cape region, South Africa.

good deal of this is natural: the fynbos, like most heathland systems, has burnt and regenerated again and again over the millennia. It is reckoned that about 340 million hectares of fynbos burn every year: fire is the natural state of fynbos, or one of them; many species need fire as part of their cycle of their lives. As a result, it is very hard to find a plant more than 20 years old. It is another aspect of the great issue of timing. Unlike some of the plants of rainforest and desert, most of these fynbos plants can't afford to take their time. This is an ecosystem in a permanent state of hurry, with the threat of fire and oblivion forcing every plant to mature and set seed relatively early in life, or pay the consequences that come with failure to become an ancestor.

This is the system the fire lily opts out of. As the annual flowering frenzy continues up above, the fire lily stays dormant down below the surface, almost as if it was sulking. It does so in the form of a bulb, a modified stem that contains the food reserves that allow the plant to survive. A bulb doesn't need to germinate; it is ready to go. It is a strategy that allows some plants to steal a march on the others: daffodils leap from bulbs at the first hint of spring while those that need to sprout from seeds must wait longer. The fire lily bulb is also waiting, but not for spring. It waits for fire.

When a fire sweeps through the fynbos, often with hundreds of species in bloom at the same time, the ground is cleared. Within four days of the fire, the first green shoots of the fire lily are visible above the ground. Their growth is triggered not by heat but by smoke: the plants respond to chemical stimuli in the smoke and they awake. Soon they are in bloom: the only flower for miles around. They are tiny, and would struggle to compete with the rest of the fynbos flora if they flowered at the same time. But they don't. In the blackened and ashy landscape they provide tiny patches of colour, and these are as effective a come-hither signal as the most extravagant bloom the fynbos plants ever produced. That's because there is no competition for pollinators at all – it is their world. The plant sets seeds, wilts and dies back, even as the rest of the fynbos plants are beginning to spout again from the seedbank that lies in the much-burned soil. The cycle continues: the fynbos regenerates, creating a multi-species environment to dazzle the eyes of any human – or bee – that should happen to pass by, while the modest flower lily retreats underground, waiting for the next time a whiff of smoke disturbs the landscape, an event that lies perhaps 20 years in the future.

As this journey through the kingdom of plants continues, it becomes increasingly convenient to describe plants as things that have will, that make decisions, that respond to stimuli, that take action, that seize an advantage. At the same time, the extent to which such things are figures of speech, convenient but non-literal ways of making an explanation is becoming decreasingly obvious. Do we really need quote-marks when we say that a plant 'decides' or a plant 'thinks'? We are moving into dangerous waters – and the plants themselves are leading us there.

The traditional notion of plants is that they are completely passive, and that they are incapable of movement, only of growth. That certainty was formulated by Aristotle in the fourth century BC, but – as a fact that seems to belong in the department of the Bleeding Obvious – it probably goes back to the dawn of human existence. It seems intuitively right because we humans can only see things through human eyes and can only operate on a human timescale. It is only within the last century that time-lapse photography has made the movement of plants vivid and comprehensible to us; in the filming of *The Green Planet*, such techniques have been taken further than ever before. We can now understand – and what is more, actually see – that a plant as homely as a daisy on a suburban lawn is a dynamic and mobile being.

The Day of the Triffids, John Wyndham's novel of 1951, is a nightmare fantasy about plants that can move and think and coordinate and of course, attack humans. The idea of a plant moving at all is weird enough to be frightening; the idea of them moving in a purposeful fashion is, in Wyndham's skilled hands, nothing less than terrifying. (The triffids are farmed for their valuable oil; the idea of an apocalypse created by the lust for oil has become still more familiar to us all in recent years.) The idea of the triffid, the plant that is coming to get you, is known to millions, most of whom have never read the book, or for that matter, even heard of it. And yet the daisy on the lawn moves with purpose and decision every day of its brief existence.

The name daisy means day's eye, and it is doubly appropriate. Daisies are shaped like the sun, which is the eye of the day ('sometimes too hot the eye of heaven shines,' said Shakespeare in the most famous of his sonnets). Daisies also open and close daily: at night they, as it were, screw their eyes up to protect themselves from cold and damp; when the next day begins they open them up again, ready to get the most they can from the sun.

OPPOSITE
Plants with purposeful movement: marguerites seeking the sun.

The opening and closing is one form of movement, but it doesn't stop there. The daisies also turn their heads during the course of the day, tracking the sun across the sky. It's a stratagem called heliotropism: turning towards the sun. At night you can see the closed flowers (technically they are inflorescences, comprising more than one individual flower) of the daisies pointing in random directions; once it is light again they operate as one, turning their faces to the sun.

There are motor cells just below the flowerhead, and they operate by changes in pressure of potassium ions within the plant. They move around a flexible segment of the plant called a pulvinus. There are two advantages that come from this in cool climates. The first is that facing the sun warms the flowerhead, and that in itself is a reward for the pollinator. On a cool day it is more agreeable and less energy-sapping to work a warm flower than a cold one: warm flowers give energy, cold flowers take it away. The second advantage is that the heat of the sun stimulates the production of pollen: the more pollen a plant can produce, the more chance it has of reproducing itself. A plant that follows the sun can offer potential pollinators more pollen more pleasantly. It's a further extension of the way that one kingdom of life both nourishes and depends on another kingdom of life, at the same time exploiting and being exploited. It is the most basic mechanism of life.

This principle holds up in the lives of arctic foxes – animals that create gardens. These gardens benefit the plant communities of their environment, and by extension everything else that lives there as well. Arctic foxes are the Capability Browns of the far north. They make their gardens without deliberate intent, but the results are of profound importance to many species of plants – and other animals. Here the time when it is warm enough for plants to grow, produce flowers and set seeds is short; for some almost impossibly short, given the poor supply of necessary nutrients in the soils of the Artic. But for those that can take advantage of the gardens created by arctic foxes, there is a much greater opportunity to become an ancestor. The soil around the dens of arctic foxes is a great deal richer than the soil elsewhere. The circumpolar regions are not only more intensely seasonal than anywhere else on Earth, they are also on such poor soils that the region could as well be considered under the desert chapter as the seasonal one. These plants face a double challenge, then: an extremely narrow window of

time in which to operate, and extremely poor soil on which to do so.

Many of these arctic plants produce seeds in immense numbers and send them out immense distances on the wind. As we have already seen, this apparently random strategy is a numbers game: the more seeds you produce, the greater the likelihood that one or two of them will land in a promising place, one with more nutrients than the average square foot of tundra. Like an arctic fox garden.

In the warmer months the foxes seek out a den. These are mainly historic structures, already dug and much used, there to be reused and perhaps enlarged, improved or repaired. The point is that plenty of foxes have been there before. A den is the heart of the foxes' existence during the lighter months of the year, and here they will raise their cubs. There can be as many as 25 pups, the biggest litter size in the order Carnivora. The area all around

ABOVE
Capability fox: Arctic fox create inadvertent life-bringing gardens on the tundra; Svalbard, Norway.

OVERLEAF
The foxes and their summer gardens: an Arctic fox cub in July; Hornstrandir Nature Reserve, Iceland.

the den, up to 20 metres in radius, is where the foxes concentrate their
energies as the cubs grow up. It follows that this is the place where the fox
family will urinate and defecate, putting nutrients into the soil, just as the
boobies do to create the forests of stunted cacti on San Pedro Martir, as we
saw in the previous chapter. The foxes also bring back food to the den: the
favoured prey of summer are the snow geese that come to the far north to
breed; lemmings are a year-round staple. The uneaten parts of the carcass
become part of the economy of the soil, and they too help to create this rich
circle of land with the foxes' lair at its centre.

These foxy gardens have usually housed generation after generation of
foxes. They can be a good hundred years old, perhaps much older; the far
Arctic, beyond easy human reach, doesn't change much. The dens are often
in a slightly elevated situation: it's a good idea to have a good view of both
potential prey and potential predators. It stands to reason that a good den in
a good place will be used year after year, and every year the foxes enrich and
nourish their inadvertent garden. The foxes are ecosystem engineers, to use
the technical term, and these places become centres of excellence, easily
visible from the air, round oases of fertility in the vast desert of the high
north. Here you find sea lyme grass and diamondleaf willow, as well as
saxifrage and Arctic gentian and other species of flowers.

Naturally these places attract plant-eating animals. Some, like caribou (or
reindeer) are beyond the scope of the foxes as prey, but they help themselves
to the bounty of the gardens, and leave their own droppings behind as they
do so. But others, like lemmings and arctic hares, are within the foxes' range
of catchable prey – so the gardens play a part as larders, the foxes feeding the
plants that feed the herbivores that feed the foxes. But there is an additional
advantage for the lemmings: when winter comes around again, the snow at
the den sites can be four times thicker than elsewhere, because of the
insulation offered by the buried plants. That makes these places important
for the lemmings that live beneath the snow. Their presence there in winter is
no immediate bonus for the foxes. By this time, they have become nomadic,
often following polar bears for what they can scavenge. But in their absence
their gardens help to nourish and maintain the lemming population:
lemmings are animals that operate on a boom-and-bust rhythm, and that
naturally affects the population of their most consistent predator, the foxes.

The foxes have a vested interest in a healthy lemming population, and their gardens, based on the bones and the droppings of centuries – many of them lemming bones – help to keep the lemmings safe. The lemmings take advantage of their own most effective predator, and often live right on top of a fox den. And here, when the growing season comes around again, the windblown seeds from the foxy gardens will sometimes find a spot in another garden, where the nutrients help them to grow, flower and release seeds into the environment: and as these are taken off by the wind, the lucky ones will find another such garden created by the arctic foxes.

The nature of life in the seasons creates the gorgeous spectacles of mass flowering. This is all very well, so far as the plant is concerned, but the business end of the growing season is the mass seeding that follows. Loosely speaking, a plant's aim is to produce seeds and become an ancestor. So after the plants have produced their flowers and competed hard for the services of

ABOVE
Window of opportunity: many species seek to flourish at the same right time in June; alpine meadow, Nordtirol, Austria.

pollinators, they must produce seeds and compete for the best spot in which to establish them. This must be done with both speed and efficiency: the Arctic winter is not far ahead, when the ground will harden and conditions become too cold for plant activity.

The first aim is to get them a reasonable distance from the parent plant. A plant wants its progeny to do well, but a perennial plant – one that has ambitions to survive for longer than the following year – finds no advantage in being outcompeted by its own progeny. The prairie violet, which is native to Canada and the United States, does so by means of a controlled explosion. This pings the seeds a short distance from the parent plant. This has a double advantage: the seed is likely to land in the same sort of habitat (or micro-habitat) that nourished the parent plant, which makes it a good place to be, while, at the same time, it won't threaten or be threatened by the parent. It is

a relatively simple and effective strategy. This same explosive technique can be found in Himalayan balsam and a plant called the squirting cucumber.

But dispersal over greater distance can be an advantage to plants. It gives them more options, more chance of finding new places where they can thrive. It helps, of course, if the seeds can home in on the same type of habitat that brought them into existence. If a plant can find an obliging being to carry the plant from, say, one area of prairie to another area of prairie, it is onto a good thing. That is why many prairie plants have evolved to exploit bison (often referred to as buffalo, but bison is technically correct). Some plants have developed hooks, spines and glue, all of which allow seeds to attach themselves to the limbs of grazing bison. Other seeds, without such specialised structures, also get entangled with the bison hairs and carried along. These seeds may well stay on some considerable time, and time for a

ABOVE
Benign explosion: Himalayan balsam expels seeds with force, so that they land a certain distance from the parent plant; Surrey, England.

bison often means distance as well. Bison are a migratory species, following the course of the seasons, moving north in search of fresh grass in the spring, and back down south again as the weather changes and the grass dies off.

You will have spotted the drawback to this strategy: the recent shortage of bison. At the beginning of the nineteenth century there were 60 million bison in North America; 100 years later there were 300 in the United States. There has been something of a recovery since then, but the system that prevailed for uncountable millennia has been destroyed. Only five per cent of the wild prairie that existed before the arrival of the Europeans is still intact, and what's left now exists in a series of islands. That makes it tough not just for the bison, but also for the plants that depend on them. It's a classic example of the way that humans have shifted the evolutionary goalposts.

There are fewer bison, but there are still plenty of tornadoes. Open grasslands are naturally windy places, and extreme weather is a fairly regular thing. You can draw a north-south line right through the middle of the United States, from northern Texas to South Dakota: an area traditionally known as Tornado Alley. Like the haboobs in the previous chapter, this different type of dramatic weather is an opportunity for plants to spread themselves. States in Tornado Alley tend to get around ten tornadoes every year for every 10,000 square miles, so it can be regarded as a freak occurrence that you can rely on. The routine tornadoes will have winds of more than 100 mph, and travel a few kilometres on a front 80 metres across, but occasional monsters produce winds of 300 mph and travel up to 100 kilometres on a front up to three kilometres wide / across. Here, then, is a matchless opportunity for seeds: sucked high into the air, to come down in random places. When there was more prairie to land on, it was a highly efficient system of dispersal. It is now less certain, but the rule remains that if you produce a lot seeds, the possibilities of landing one in a prime spot increase drastically. The chances of becoming an ancestor have diminished in the last couple of centuries, but the plants are still playing the percentages as best they can.

If this sounds as if I am attributing conscious calculation and intelligence to plants, I am being misleading. But the temptation to talk about plants as thoughtful and intelligent beings can be overwhelming when you encounter a plant with as smart a lifestyle as a grass species called *Ceratocaryum argenteum*.

You can find this in the fynbos area of South Africa, which has already been considered in the context of the fire lily earlier in this chapter. The area is heavily seasonal, as we have seen, and prone to fire. *Ceratocaryum* flowers in the high summer and drops its seeds around the time that the rains begin. Timing, as always, is a big deal in the seasonal world, but the reasons for the timing in this case are unusual. It involves one of most highly evolved pieces of deception that you can find anywhere in nature. The plant drops its seeds at the time when the dung beetles are most active. It does so in order to dupe the beetles into burying the seeds.

Dung beetles have always fascinated humans: the way they push a ball of dung, often bigger than themselves, across difficult and uneven going before burying it. Into its warm, decaying mass they will lay an egg, behaviour which seems to us to encapsulate all kinds of admirable traits: self-sacrifice, parental care and the dignity of labour.

ABOVE
Fooled again: this is not the antelope dung a dung beetle needs, but a seed that mimics the dung to exploit the beetles' dung-burying behaviour; De Hoop Nature Reserve, South Africa.

The *Ceratocaryum* exploits these traits for its own purpose. Each plant will drop about 50 seeds, every one of which looks like a pellet of dung – dung that might have been dropped by the antelopes that live in this part of the fynbos region, elands and bonteboks. What's more the seeds smell like dung. This double deception is enough to bring in two species of dung beetles. Not only do the seeds look and smell like dung, they are, temptingly, more spherical than usual, and so easier to roll. Each one is about a centimetre in diameter and for a dung beetle, it is the most desirable object in the world.

So they roll it away and bury it. The smaller of two species of beetle involved will bury the seeds one at time, which is a better deal for the plant; the larger one will do so in clusters of ten or more. Neither beetle will take them very far from the plant, but that is not a problem. The seeds, like so many in the fynbos system, won't germinate until there has been a fire, which will have destroyed the parent plant – leaving a great opportunity for the emerging sprouts of *Ceratocaryum*, which come up after the time of the fire lilies. As for the beetles, they realise that they have been duped when they attempt to eat or lay eggs in the object they have so laboriously buried. When they come to that realisation, they emerge looking for another opportunity: and will as like as not home in another *Ceratocaryum* seed. So desirable are these seeds that the beetle will fight for them, battling for the right to be duped.

Once a plant has set seeds, its job is done. At least for now. Some plants sprout, wither and die in the space of single short season, leaving only their seeds behind, in the hope that a subsequent generation will also live a short but productive life: one that can be measured in weeks. Such plants have no reason to fear the winter: they are not even there, save in the form of buried seeds. Plants that take a more long-term view need to find a different strategy: for in the intensely seasonal parts of the world the winter offers challenges that seem impossible to overcome.

How extraordinarily, how impossibly different is a summer landscape from a winter landscape, in the countries of the northern hemisphere where the seasons dictate the pace. Even a landscape of arable fields is different: the green and then the yellow wheat is gone, replaced by bare earth, last year's stubble or the tiny spouts of winter-sown crops, often white with frost, at times wrapped in snow. The remaining woodlands in these countries must

OPPOSITE
Forever changing: autumn or fall in deciduous forest, Acadia National Park, Maine, USA.

OVERLEAF
First snows: October in the highlands of Maine, USA.

change greatly... as greatly, perhaps as Superman when he resumes his identity as Clark Kent.

For people unused to the rhythm of the seasons, the changes are so massive as to seem unreasonable. Winter is a crisis that occurs every year, and every tree must find a way of coping with it or die. There was a Cold War term for the feared deployment of nuclear weapons: crisis relocation. It means being somewhere else when the disaster happens. That is the option taken by many species of birds, the short- and the long-distance migrants. By the time winter has come to their lands of summer plenty, they are somewhere else.

It's a great option, but it requires wings. Some small mammals find a place to hole up and hide. There they sleep the winter away, living on fat reserves, only waking from hibernation when the sun and the warmth return. Trees can't fly, trees can't hide – and yet in the winter time trees can't feed. They can't survive as they are, and so they become something else.

Trees get their lives from their leaves. The leaves allow them to photosynthesise: to make food. But in winter these life-givers become liabilities. Hard weather would cause the moisture inside the leaves to freeze, expand and rupture their own cell walls. Even if that weren't the case, the tree would seldom be able to find enough energy from the winter sun and the season's brief days to make the business of photosynthesis possible. The evergreen trees, the conifers, can cope with all this, because their tiny leaves – needles – are covered by resin, a waxy overcoat that protects them against water loss and freezing. (Some of these conifers – about 300 million across the world – will be taken into houses as symbols of the persistence of life through the crisis of winter, as Christmas trees.)

But the non-evergreen trees – those with leaves so perfectly equipped for gathering light in the long, light-filled days of summer, those plants that so perfectly demonstrate the principle of maximal outside – must find some other way of coping. So they more or less turn themselves upside down. They withdraw all the good things from the leaves and send them down to the roots, the exact opposite of what happened in spring. Sap, nutrients, water, sugars: all these things go underground. Sap that rose in the spring and in America nourished the sapsuckers now sinks back down again.

Chlorophyll is the molecule that absorbs the sun's energy and allows

OPPOSITE
The annual death: American oak leaves go through the colour changes before they fall.

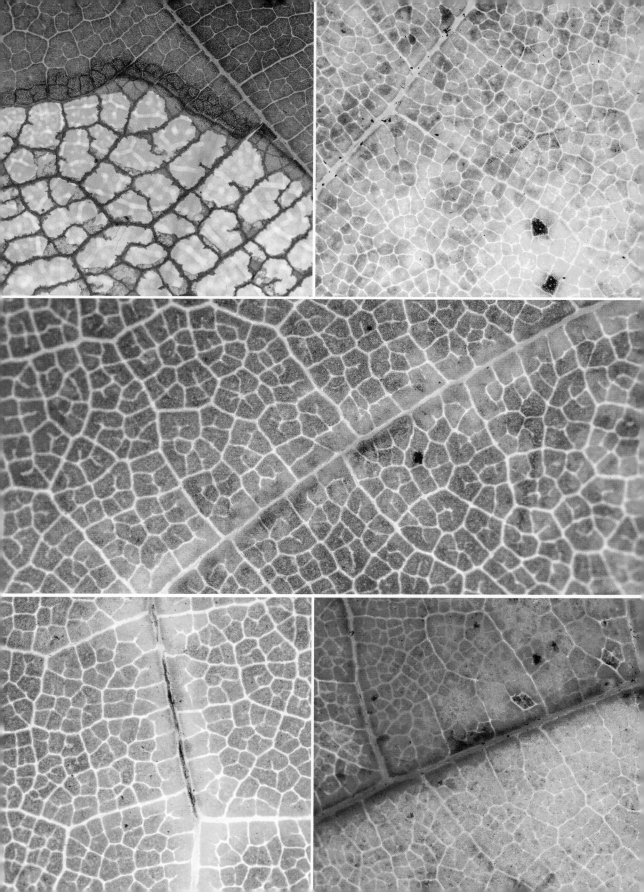

photosynthesis to take place. And of course, it's green. Chlorophyll is one of the first molecules to be broken down by the tree as it prepares for winter. It does this so that it can reabsorb its own nutrients, rather than waste them. It follows that the trees, or rather their leaves, stop being green. As the chlorophyll moves out of the leaves, the departure of the green reveals the yellows and oranges that that lay beneath. This complex process creates an annual spectacle of death and the promise of renewal.

As this process continues, every tree sends a chemical message to each leaf, causing the cells at the base of each leaf to weaken their cell walls while other cells expand. This creates an effect that has been compared to perforated paper: an invitation to tear. In this way the leaves are nudged from their twigs, so the leaves do not so much fall as get pushed. Perhaps it would be more correct to refer to autumn as the push, rather than the fall.

There are other advantages that come with this leafless state. The first is that without the great sails of leaves, a tree is less vulnerable to the wind, and so less likely to get blown over. When the spring comes again, some trees put out flowers before they have grown leaves. For windblown pollination there is an advantage, as there are no leaves to get in the way of flying pollen; for those plants pollinated by insects and other animals, it is easier for the pollinators to find the flowers.

The loss of leaves is called abscission. The trees stand naked and unable to feed. They must hang on, living on their stored resources, until the seasons change again, the days get longer and warmer and world is full of light. Then the sap will rise again, the trees will once again grow leaves, and these will trap and eat the light that floods down for hour after hour, day after day – until the Earth tips back again and the cycle of the seasons continues.

There is a temptation, when describing the vast and extraordinary mechanisms of nature, to write as if everything was honed to perfection and nothing could ever go wrong; that nature is a perfect steady-state perpetual motion machine that lasts forever. That is not the case. Changes happen all the time, things beyond the control of the individual organism. Changing conditions are disasters for some species but they can sometimes be glorious opportunities for others. We have talked about perfection of timing but, in the seasonal systems that depend on timing, there will be occasions when there are temporary or permanent alterations to the seasonal pattern, or when an organism simply gets it a bit wrong – misses the moment and fails to find the window. Just as there have been times when the apparently perfect athlete Usain Bolt got his timing wrong – at the World Athletics Championships in 2011 he was disqualified in the men's 100 metres final for false starting – so there are times when conditions catch out the most accomplished exploiters of the seasonal niche. I once saw a swallow in England in late November: it had missed the moment to migrate to Africa. The annual winter shortage of flying insects – swallow food – would surely have killed this reckless long-stayer soon after.

Plants – wild plants, not just delicate favourites cultivated by optimistic gardeners – can get caught out by a sudden and dramatic shift in temperature. In late autumn and early winter, in the strongly seasonal parts of the world, there can be occasions the air is suddenly and drastically cold, well below freezing, while the soil is still cosy and warm. In this event a plant, with its roots secure in the soil, tries to continue with the processes of life, but the weather up top makes such an attempt futile and self-destructive.

In these circumstances you can witness the creation of ice flowers, sometimes called frost flowers. They are not flowers at all: rather, they represent a beautiful rupture, like a poetic deathbed speech by a

ABOVE
Beautiful disaster: an ice-flower bursts the stem of this Verbesina virginica *in an autumn frost; Mingo Swamp, Missouri, USA.*

Shakespearian hero. The sap within the plant above ground freezes – showing us precisely why the trees are so keen to lose their leaves – and it expands, because water expands when it becomes ice. This causes the plant to form cracks along the stem. Water is drawn to these cracks by the capillary action of the plants, a process we have already looked at in the first chapter. But in this instance the water freezes when it makes contact with the air. More water comes in behind this, pushing the ice out from the stem. These form beguiling shapes that look like flowers, a macabre and fleeting beauty that reminds me of a line by Bob Dylan: 'The flowers of the city, though breath-like get death-like sometimes.'

Throughout this chapter we have been applauding the plants of the seasonal world for their mastery of timing. Here is a small, beautiful and disturbing example of the sort of thing that can happen when their timing is a little bit out, when they fail to judge the seasons to perfection, when they are undone by a blip in the transition of the seasons. Things can also go wrong when there is a wholesale alteration of the seasonal pattern – precisely what is

happening all over the world. We will look at that subject again in a few pages: remember the frost flowers, and how they show us what happens when plants and seasons get out of step.

We have seen how trees wake up in the spring, how the sap rises and many other living things take advantage. The reverse process is also important for life beyond the trees themselves. As the sap sinks down to the roots, it creates an opportunity for a quite different group of living things: fungi. Autumn is the time when any woodland walker will see a great proliferation of fungi, growing from the trunk, making rings around the roots, appearing and disappearing in their distressingly sudden manner. We notice them once or twice; the next time we pass that way they are gone. We are entitled to wonder about this curious way of life: here today, gone tomorrow.

But the fact is that fungi are there all the time, and they are essential to the lives of the trees. A mushroom or toadstool is not the entire organism: it

OPPOSITE
Underground partnership: ceps or porcini mushroom in the New Forest National Park, England.

ABOVE
Sinister perfection: fly agaric fungus in coniferous woodland near Inverness, in the highlands of Scotland.

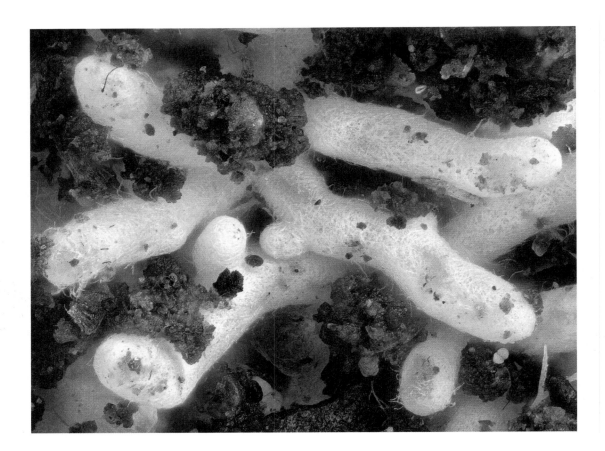

can be compared to the fruit of a tree. You don't mistake an apple for an apple tree or an acorn for an oak: and in the same way, the visible part of the fungi is only a small part of a great story.

It is traditional to suppose that fungi are the enemy of trees, leaching their energy and bringing in diseases: the fewer fungi, the better it is for the wood. The exact opposite turns out to be the truth. Without fungi any wood would be less than itself, perhaps even unable to exist at all.

To understand this we must do away with our traditional idea of what a fungus is: a mushroom in a plastic box in the supermarket ready to become part of a risotto, or a spotted toadstool full of deadly poison. The fungi we see, in a wood or on a plate, are the fruiting bodies, there to pump out their spores into the air and help the fungus to become an ancestor. The rest of the fungus – the fungus itself – is a thread. These hidden threads are the heart and the

ABOVE
The fungal self: threads of mycelium on the roots of pine trees, making one of the most powerful relationships in nature.

soul of a fungus, what it does and what it is for. The bits we see are more or less afterthoughts. The threads are called mycelium and the first thing to understand about them is that they exist in quantities impossible to understand. Many are far too fine to see, quite invisible to the naked eye. You can find 600 metres of mycelium in a single gram – say a teaspoon – of soil. Take a nice big handful of the stuff and you will have thousands of metres of mycelium. This is life in a form that exists beyond our easy understanding, and it reaches out through the soil and taps into the roots of trees.

As it does so it establishes one of the most extraordinary relationships in all nature, one that works to the benefit of both participants. Fungi, as we said before, are not plants, and are more closely related to us animals – like us, fungi are consumers, while plants are the creators of food. The fungi are able to get the carbohydrates they consume from the plants that create it – but at the same time the plant takes water from the fungi and, with it, nutrients like nitrogen and phosphorous. The fungi get these from the soil, and are able to pass them on because they possess enzymes that the plants don't. Both parties are richer for the association. Perhaps each party makes the other's existence possible.

The network of mycelium beneath the surface of the soil stretches out in all directions. It links one tree with the next, sometimes of the same species, sometimes quite different species. The threads among the roots of the trees join tree after tree, and do so right across the forest. The trees can't reach out and touch each other, but they can communicate at one remove, by means of the network of fungus. It's as if every tree in the forest was in the lockdown state we all experienced during Covid-19, but is engaged in a perpetual Zoom meeting that involves every tree in the wood – and what's more, it doesn't mysteriously lose connectivity when you get to the crucial item on the agenda.

By this means, big successful trees that have spread out their leaves in the canopy can transfer some of their nutrients to plants in suboptimal conditions – those nearby that are not tall enough to access the direct sunlight as a canopy tree does. It is like the reverse of parasitism: a plant not taking things from another but being given them. The nature of this apparent altruism is baffling but, certainly, small plants that have germinated in the shade of a giant are likely to be the offspring of that giant. Does the giant preferentially and knowingly seek out its own progeny for such lavish gifts? It's a question to

make the head spin, but then everything to do with the fungal network is capable of doing that.

Certainly every wood in the world benefits from its own extensiveness: the bigger it is, the safer each individual tree is from major weather events; the further a tree is from the forest edge, the safer it is. A tree that helps the forest to get bigger is looking after its own interests.

When one tree in the forest is affected by disease, the information about its condition is communicated to other trees by way of the fungal network. That allows the unaffected trees to boost their immune systems. Information about a disease triggers a tree into producing defence-related chemicals. It is a pre-emptive medication, a little like a vaccination in spirit. It means that, should the disease strike, each tree that's connected with the fungal network will be quicker and more efficient at coping with it.

ABOVE
Family of giants: trunks of giant sequoia with some smaller individuals behind; Sequoia National Park, California, USA.

The communication system also works when it comes to dealing with damaging animals: what a forester or a tree would regard as pests. It's been demonstrated in laboratory conditions that broad bean plants that are linked by a fungal network can inform each other when an individual has been attacked by aphids. Neighbouring plants respond by producing chemicals that protect the plants from the aphids: should they arrive, the plant is ready for them.

But there is a chilling side to this extraordinary system of communication. Plants will sometimes send toxins through the fungal network to a competing neighbour, making it hard if not impossible for a plant to establish itself. If there is a moral here, it is that nature doesn't exist to teach moral lessons to humans. If we seek moral lessons then we must learn them from creatures capable of moral choice.

The lesson that we really need to learn from the discovery of the wood-wide-web, and our early – and still rather coarse – understanding of it, is about biodiversity: the way it takes many different species to create an ecosystem, affecting each other in ways that are often difficult to understand, let along calculate. Fungi need trees, but then trees need fungi. In diversity lies strength and resilience. Vast and mighty living things often depend on tiny and apparently insignificant things for their continued existence: that is part of the great pattern of biodiversity. So let us turn to some of the biggest, oldest and toughest living things that have ever existed on this planet, and see how they are faring.

The sequoia trees of California are evergreens. They keep their leaves – needles – throughout the winter, and this is the time that they flower and disperse their seeds. They were once found in many other places, but the climate changes quite naturally over the course of the millennia and they are now reduced to a population in a small corner of California. They are not quite the tallest trees in the world – that honour goes to their near relatives among the redwoods – but they are the most massive. They are the biggest living individuals on the planet (the Great Barrier Reef is visible from space, but it comprises uncountable millions of individuals) and quite probably the biggest there has ever been.

That means, obviously enough, that a mature sequoia requires a lot of water. A 90-metre giant needs 2,000 litres of water a day, and is better off with 4,000. Resources on this scale can't be guaranteed every day, and to survive for these immense lengths of times – 3,000 years is well within the range of these trees – they require qualities of tolerance and resilience. They are able to take water from reserves in the ground, and to bide their time during periodic droughts until the times of plenty are back. You don't reach that sort of size without enduring changes and the difficulties they bring.

Or, to put that another way, it takes a lot to upset a sequoia. They've seen it all, they've endured the bad times and come through to the good times again and again. But now, and there is no ducking the matter, the sequoias are upset. That can only be because a lot has happened to upset them. They are showing this distress by dropping needles and even branches. The deciduous trees drop leaves as a way of husbanding their resources, so they can get through the entirely predictable hard times of every winter. An evergreen tree also drops its

OPPOSITE
Visiting a cathedral: David Attenborough in Sequoia National Park, California, USA.

leaves – its needles – as a way of self-protection, but not as part of the normal rhythm. This is a response to the exceptional.

The shedding of leaves is an effort to conserve water. As a Christmas tree sheds its needles across the sitting-room as it dies in front of the family that brought it inside, so the sequoias are shedding needles in an effort to stay alive. They are accustomed to getting their water from the melting snow of the Sierra Nevada, the 400-mile-long mountain range that lies mostly within the state of California. These mountains have always been, so far as the trees are concerned, a great slow-release water storage system, the annual snow gradually melting throughout the year and working its way down to where the trees can access it. But now much less snow is falling, and it is melting much faster when it does. The water stored beneath the surface is depleted. Measurements taken over time show that the sequoias are now responding less well to drought than they did even a decade ago. All this is combined with other problems, to do with the destruction of much wild habitat all

ABOVE
Green Planet *team checking footage in Sequoia National Park, California, USA. Left to right: Rosie Thomas, producer, Rupert Barrington, series producer, Robin Cox, cameraman, David Attenborough, Alistair Tones, assistant producer.*

around, and the suppression of the natural cycle of forest fires. But the problem that has upset the sequoias is climate change.

We have been playing fast and loose with the unimaginably complex system that operates the planet. We are now beginning to learn about the unintended consequences of our actions and decisions. For two decades and more, scientists and conservationists have been pointing out with increasing urgency that this is serious, this is an emergency, this is really something that matters. David Attenborough described climate change as 'our greatest threat in thousands of years': that is to say, bigger than the Black Death, bigger than world war. The great struggle is for us humans to find a way of understanding that it is happening now and that it really matters. The vastest trees of the world, trees that have endured for thousands of years, the toughest things that have ever lived, are shedding branches because they can't cope with what we've done. Does that help us to grasp the reality of this crisis?

ABOVE
Sky-reaching: a giant sequoia in Sequoia National Forest,
California, USA.

WATER
WORLDS

Life began in the sea and moved onto land. We like to portray this transition in the most heroic terms: as the conquest of land, as if our ancestors had to fight some fearful battle, one that would eventually allow us humans to live as we do. We generally forget that the process began with the plants. Well, of course it did. There would be no point in animals or fungi moving onto the land in a big way unless there were plants there to live off. It was a long and complex process, but fossilised remains of mosses and liverworts have been found dating back 470 million years. Such plants were followed by animals, at first small invertebrates, and eventually – a date estimated with unnerving precision at 367.5 million years ago – the first backboned creatures, the tetrapods, were walking on the surface of the Earth.

But all this heroic march-of-progress stuff gets muddled. Some land creatures started to return to the water, showing us the confusing reality that evolution is not a straight-line business and that it doesn't operate with the specific goal of producing humans. Life will take the form that conditions dictate: if there is a possibility of making a living in a way that seems regressive, the chances are that some kind of living thing will take it. We have already seen, with parasitic plants like the desert mistletoe in Chapter 2, which evolution is perfectly prepared to abandon complexity in favour of simplicity when simplicity gives an advantage. We have also seen, again and again, that creatures who evolved for the land moved back into the water and made an excellent living by doing so.

We mammals evolved for the land, but there are now many water mammals, including otters that retain their legs and can operate as well on land as they do in water, seals with legs modified into flippers, but who must still give birth on land, and whales and dolphins who are completely aquatic and have as little as possible to do with land.

This same story of reverse evolution, a return to the water, can also be found in the kingdom of plants. The stories here are less well known and much less obvious but, all the same, plants that evolved for the land have produced species that now make water their homes, and the advantages and challenges that they face are as enthralling and, if you like, as heroic, as those of any plants that prefer to keep their feet dry.

In some ways the plant that returns to water has a lot going for it, the first and most obvious advantage being continuous access to water. Water plants

OPPOSITE
Salad days: a water vole feeding on aquatic plants; Kent, UK.

live lives remote from those of cacti and the other desert plants in Chapter 2: no need to hoard and protect water when it's with you in immense quantities for every second of your existence. Many moving waters are rich in oxygen and full of nutrients, so there is no need for fantastic adaptations to acquire these essentials. A plant on the surface in open water is well away from the shade of the trees: the competition for light, which we witnessed in the rainforest in Chapter 1, doesn't exist in the same way for plants that have returned to the water, although, even here, depending on the depth of the water, plants at the surface can eclipse those below.

Every environment provides different challenges, and different forms of life adapt to cope with them in different ways. A fast-moving stream is one of the most challenging environments on the planet for a plant. Water – like precious jewels for a cactus – goes rushing past a plant in a mountain stream,

ABOVE
Holding fast: flowering Apinagia *in the Peti rapids,*
Gran Rio, Suriname.

thousands of litres every hour. The problem is that the water is likely to rip a plant out by the roots and send it cascading downstream. Mountain rivers are very destructive, collecting rocks and stones and throwing them about, and any plant that can't hold tight is a goner. They must survive immense forces, holding fast in conditions that would knock a human sideways.

We talk about 'pure as a mountain stream', and it's a good analogy. Such streams are so pure they contain very few nutrients. This is another challenge for a water plant: they must collect nutrients that come in tiny quantities in immense amounts of water. Then there is the question of light. Water stops light far more readily than air. Any underwater photographer will tell you that 'water eats light'. The deeper a plant lies beneath the surface, the less light is available. When there is sediment, the problems are greater, but all plants that live in water must cope with the fact that the deeper their leaves are beneath the surface, the harder it is for them to photosynthesise – to make food. To add to these difficulties, water levels vary, and often do so from one time of year to the next. A plant growing at the perfect height for its continued existence in one month might find itself deep below the surface later in the year, and at other times exposed to the air. In such conditions versatility is essential to survival.

This principle is demonstrated to an almost fantastic degree in a 100-kilometre stretch of river in Colombia. It is called the Caño Cristales, a tributary of the Guayabero River, which is part of the Orinoco river system. Since the civil troubles in Colombia died down and it became safe to travel in the country again, the Caño Cristales has been recognised as one of the great spectacles of the natural world, and has acquired many nicknames: Rainbow River, the Liquid Rainbow, the River of Five Colours, the river that escaped from Heaven or simply the most beautiful river in the world.

It flows fast, with many rapids and cataracts. Its level varies from season to season. The dominant plant in these waters is *Macarenia clavigera*, and its most obvious challenge is to hold tight to the rocks throughout the year. When the optimal conditions occur, though, everything changes and the plants come into flower. When there is precisely the right amount of light and the right temperature, and the water drops to the right level and it flows good and blue over black rock and yellow sand, the plants show both green and an astonishing shade of red, perhaps the last colour would expect to see on a river, at least in great masses. For a few brief weeks this combination creates one of the most glorious and unexpected shows in all nature. The normal and predictable colours of riverine life are subverted: here is a riverscape of yellow, green, blue, black and blinding red.

By exploiting this thrillingly difficult, swift-moving environment, these plants are ecosystem engineers: they make life possible for many other species. The invertebrates and fish of the river depend on the nutrients that these plants store in their tissues. The plants' ability to hold on tight in appallingly difficult conditions means that these rivers are home to many other species.

In the face of such an extraordinary detonation of beauty, it seems carping, almost wilfully miserable, to suggest that such a spectacle may not be with us for much longer. But it is the duty of conservationists to look at the

OPPOSITE
The liquid rainbow in full flow: two shots of the Caño Cristales,
La Macarena, Colombia.

world as it is, not as we would like it to be, so let us note in passing that the river that escaped from Heaven is under threat by oil exploration and by land-clearance projects upstream.

You'd think that the cleanest, freshest and purest water would make a heaven on Earth for any plant that ever grew. But water is only one part of a plant's requirements. A plant needs minerals and nutrients if it is to function. Among other things, a plant needs nitrogen, which helps to make the chlorophyll and so makes photosynthesis possible; phosphorous, which helps with the growth of roots and flowers; potassium, which adds strength and helps early growth and water retention; magnesium which helps with the green colouration essential for photosynthesis; sulphur, which helps the plant to resist disease and to make seeds; and calcium, which helps the growth and development of cell walls. Other nutrients, often in the form of trace minerals, are also important. In normal circumstances a plant gets these vital things from the soil; that's why humans managing the growth of domesticated plants must add nutrients to the soil if they are to get repeated success. This can be added in the form of natural fertilisers like manure and bone meal, and in the many synthetic fertilisers.

But wild plants must find their own nutrients, and in some circumstances that is a serious struggle. Let us move, then, to the landscapes of the tepuis of Venezuela and Guyana: the vertiginous table-lands, the elevated plateaus that inspired Sir Arthur Conan Doyle's novel *The Lost World*. There are, alas, no dinosaurs to be found on the real tepuis, but if you look closely, there is the stuff of a fantasy every bit as luxurious.

It rains on these plateaus. It rains heavily and it rains daily. The rain washes away soil which is already low in nutrients; in a world of water, the plants that live there must cope with challenges every bit as drastic as those that live in the desert. It is easy to get water, but very hard to hold on to it, because there is so little soil to retain it. Bromeliads grow upright clusters of leaves. The most obvious example is the pineapple pool. You can grow the top of a decapitated pineapple; if you do so, you will note that, when you water it, it collects some of the water in its leaves, which make a little pool. There are around 3,500 species of bromeliads, most of them in the tropical Americas; some of them grow as epiphytes on the branches of rainforest trees and provide pools far from the ground, much sought by tree frogs.

OPPOSITE
Dealing with a food shortage: pitcher plants and bromeliads on a rain-washed tepui, Canaima National Park, Bolivar, Venezuela.

The bromeliad species on the tepuis also have these spikey rosettes, and they collect rainwater. The small pools that result are attractive to insects. The leaves of these plants are often bright yellow in colour, very conspicuous. Insects are tempted to examine these pseudo-flowers; often they slip on the waxy surface of the leaves, slide into the pond and drown. Some invertebrates are capable of using the pools without dying; these bring nutrients in the form of their minute droppings. Thus the bromeliad has brewed a nutrient-rich soup in its cup of leaves, and these leaves absorb its goodness.

The bromeliad pools are a rich source of food in a nutrient-poor place. That makes them attractive to other living things. These include the bladderwort, which can claim the distinction of being the fastest plant on Earth; in fact, it performs one of the swiftest actions in all of nature. It is a plant that defines itself by its movement, a contradiction we have already seen elsewhere in these pages. The first part of the movement is exploration: in the long, sensitive tendrils that reach out from the plant and explore neighbouring bromeliads. When they have found a good one, they grow into the pool in the centre of the bromeliad. Once there, the tendril changes shape. It is no longer a reconnaissance device: it is a trap.

The plants grow a collection of small bladders. When they are established, glands within the plant extract the water that has shaped it. This creates a partial vacuum. The bladder is equipped with an entrance, and this is fringed with bristles. An invertebrate, living in the pond or seeking to escape from it, touches one of the bristles, and bang – the vacuum is released, the plant implodes and the invertebrate is sucked in. It's a bit like a passage in the book *Goldfinger*, in which James Bond's enemy Oddjob is sucked out of a plane window by the pressure difference inside and outside. The bladderwort then resets its trap while the plant slowly digests its meal, absorbing the nutrients offered.

Plants that catch, kill and consume animals have always haunted the human imagination. Sometimes we make them up, like the plants in John Wyndham's *The Day of the Triffids* and Audrey II in the musical comedy *Little Shop of Horrors*. But sometimes they are real, even if they don't exist on a scale large enough to hunt down and devour humans. Charles Darwin said that the Venus flytrap was 'one of the most wonderful plants in the world', and he offered his specimens hard-boiled eggs and beef to see how they would cope.

OPPOSITE
Fastest plants on Earth

TOP LEFT *Bladder of bladderwort though scanning electron microscope.*

TOP RIGHT *Greater bladderwort with trapped mosquito larva,
the protruding head just visible.*

BOTTOM *Entrance to a bladder.*

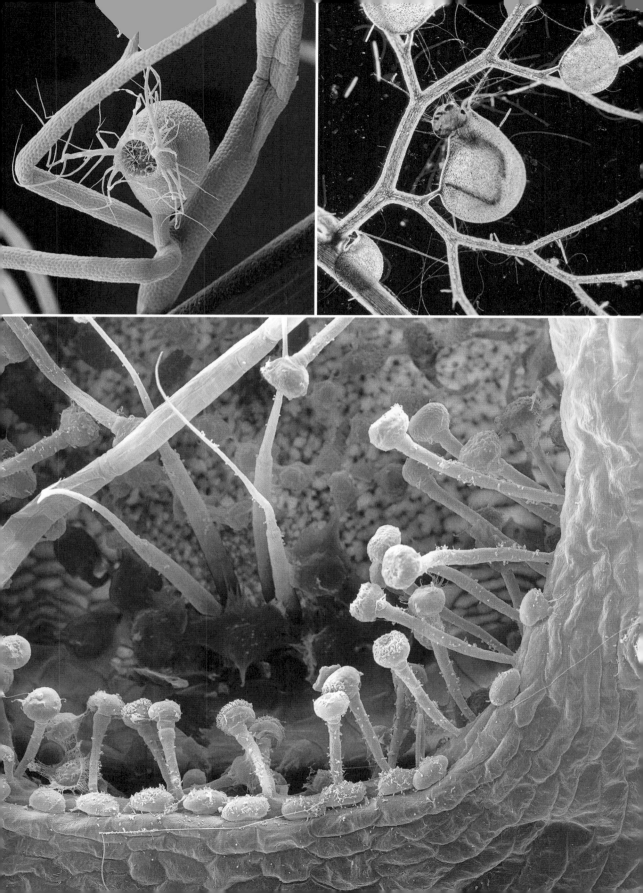

The image of the plant with spiky traps is familiar all over the world, quite an achievement for a small plant that only grows wild in the swamps and wet savannahs of North and South Carolina in the United States, and reaches a maximum height of 10 centimetres. As we have seen on the tepuis of South America, wetlands can be challenging habitats for plants. The waterlogged ground inhibits the microbial process of breaking down organic matter which in less stressful habitats makes nutrients available to plants to gather in through their roots, while flowing water washes away such nutrients as are available.

The option of becoming a carnivore – a consumer of animals – is one solution to this. It's such a sound idea that it has evolved quite separately at least six times among plants; there are around 600 species of carnivorous plants in the world. These include the sundew, which is found in British bogs,

ABOVE
The plant that knows how to count: David Attenborough and cameraman Robin Cox in a glasshouse at Kew Gardens, London, filming the Venus flytrap.

and pitcher plants, which we have already looked at in the first chapter. But there are two that have evolved snap-traps: one is the waterwheel plant and the other is the infamous Venus flytrap. The latter can close a trap on an insect in half a second; the time is variable, depending on the local conditions and the health of the plant.

Closing the trap requires a certain amount of energy, and so does opening it again. The process of reopening also takes time. It follows that the plant doesn't want to waste its time and its resources by snapping shut on stuff it can't digest. In order to function effectively, it needs, at some level, to be aware of what it's doing, to make calculations and to make a series of decisions, precisely the sorts of things the triffids did in Wyndham's apocalyptic masterpiece. The Venus flytrap is capable of all these things.

The abilities of the Venus flytrap are demonstrated in a bravura piece of footage filmed for the televised version of *The Green Planet*. The sequence shows David Attenborough's delight in magicianship: not his own powers of magic but the mind-boggling nature of the natural world. Attenborough is not a magician; he is much better than that. He is the person who allows nature to show us what it does best. At the Royal Botanic Gardens at Kew, he asked a question: how does a Venus flytrap distinguish between a flying insect and a speck of torn leaf, a mote of dust or a floating seed? He used a paintbrush to trigger a sensitive hair inside the trap – and can you guess what happened? The answer is cleverer than you might believe: absolutely nothing. But when he touched another hair within 20 seconds of the first – *then* the trap was sprung and the jaws closed.

Something that touches a single hair is probably inert: just a piece of floating rubbish. But a thing that follows up the first touch by touching a second hair – that's something that can move of its own volition. That's not rubbish, that's food. So snap, close, and thank you very much.

But digestion itself is also a considerable expenditure of energy. There's no point in digesting unless the prey is big enough to make the effort worthwhile. That's why the trap only half-closes. The spikes that, to us fanciful humans, look like cruel devices to catch and maim, are in fact instruments of mercy – or, at least, items that enable the plant to operate in its own best interest by releasing unwanted prey. The spikes form the bars of the cage, and if the creature is small enough to get out between those bars,

then so be it: it's not worth the trouble of eating, and so it goes free. Larger creatures remain trapped inside. Such prey will endeavour to escape, of course: and in doing so they will touch more of those trigger-hairs. When a prey item has touched the trigger hairs five times, the plant is satisfied that it has living prey worth eating – and so the trap closes tightly, the spines interlock, a hermetic seal is formed and the trap becomes a stomach. The digestive juices get to work, and the prey is consumed. The plant absorbs its nutrients, and so, prey item by prey item, it gets what it needs to flower and so to reproduce itself: to make more Venus flytraps.

What we have, then, is not just a predatory plant, but one that can remember and count. It's a plant that challenges the human sense of order: the *scala naturae*, described by Aristotle but no doubt an idea as old as humanity, in which plants are ranked above rocks but below animals – with humans on top, just below angels, archangels and God.

ABOVE
Too late: fly caught in a closing Venus flytrap.

OPPOSITE
No way out: wasp caught by Venus flytrap

Linnaeus, the great eighteenth-century cataloguer of nature – the first edition of his great work *Systema Naturae* was published in 1735 – rejected the entire idea of the Venus flytrap as a carnivore because it violated the principles of the *scala naturae*, suggesting instead that the plant kindly sheltered insects from the rain. But Darwin, more than a century later, was fascinated by everything that challenged our deepest assumptions. He loved the Venus flytrap and published a book, *Insectivorous Plants*, in 1875, 16 years after *On the Origin of Species*, one that, more or less incidentally, reinforces the inescapable truths he had already shown us.

But the Venus flytrap still has a problem. It needs to be pollinated, and it needs insects to do the job. To devour its own potential pollinators would be self-destructive. And since the plant comes into flower while the traps are active, this could become a self-made and self-damaging trap. The answer seems to be that the flowers are separated from the traps in space. The flowers are at the top of the plant and attract flying insects, especially bees and beetles; the traps are lower down and mostly catch creatures that get

there by crawling. The catch comprises one-third ants and one-third spiders; the rest is made up of grasshoppers, some other beetles and others – including small frogs. It's also possible that the smells and colours that flowers and traps use to draw in the animals they exploit are different, attracting different species.

The Venus flytrap looks like magic, and David Attenborough delighted in showing off its apparently magical qualities. But there is deeper magic involved. The Venus flytrap is like a favourite book, one so wonderful that it might seem to you, its most devoted reader, to be almost a thing of magic. But when you start to think about it, you realise that the greater magic is that any book exists at all, that humans have brains that create language and from that language form stories and ideas and songs. Your favourite book may indeed be a thing of wonder, but in a sense every book that ever got itself read is wonderful enough.

The same thing is true of plants. The Venus flytrap is wonderful all right, and it forces us to reconsider all kinds of traditional ideas of what plants can and can't do. But there's a still greater thing true of the Venus flytrap that's also true of the humblest fleck of living green that has ever existed on our planet: they can live. They can respire. They can make more of their own kind, so that in a sense they never die. Compared to that miracle, a plant that counts, remembers and eats spiders is, well, just about the sort of thing you'd expect, given a few hundred million years of evolution.

Plants take in carbon dioxide and give out oxygen when they photosynthesise during the day. We animals do the exact opposite. That truth is so perfect – so magical, if you like – that it might tempt us to join Professor Pangloss, the character in Voltaire's novel *Candide*, who believes that all is for the best in the best of all possible worlds: that our ears and noses are cunningly put together so that we can wear spectacles.

The truth is a little different and, if you like, a little more marvellous. It's generally accepted that life began about 3.8 billion years ago, and continued in the form of mostly single-celled blobs until about 900 million years ago. If that sounds a little unenterprising, think again. These years were essential to the biggest event in the story of life on Earth, apart from the beginning of life itself. This was, of course, photosynthesis. Cyanobacteria acquired the ability to turn light into food – and, as a by-product, they gave out oxygen.

OPPOSITE
Safe: two hoverflies on a Venus flytrap; the flower puts them in no danger.

Over the millions of years they changed the atmosphere of the entire planet. Inevitably, organisms evolved to take advantage of this new and fantastic resource. Plants developed and gave out still more oxygen: animals followed them onto the land to take advantage of it. The arrival on land of photosynthesising plants changed everything. Plants didn't – plants don't – just provide us with food. They also provide us with the stuff we need if we are to stay alive for more than a minute. Oxygen allows us to build new cells, to find energy, to drive an effective immune system. All across the planet, the oxygen that we depend on is provided by plants. You don't have to take my word for this. The televised version of *The Green Planet* is able to show us a conjuring trick every bit as vivid as Attenborough's paintbrush trick with the Venus flytrap.

So let us look at a small stream in Brazil. As said before, water can be a difficult environment for a plant, because plants need light and water stops light much more quickly than air. Below 100 metres very little light gets through at all, certainly not enough for photosynthesis to take place. Many watery environments are thick with sediment, making it even harder for light to penetrate.

But this Brazilian stream is a marvel of clarity: gin-clear with the bottom sharply visible. It is also fed by springs that rise from the bed of the stream, filling it with water rich in dissolved carbon dioxide. For the plants that live in this stream, then, these are optimal conditions for photosynthesis. It follows that, when the sunlight reaches the leaves of these water plants, a super-charged photosynthesis can take place. A view through a microscope reveals almost frenzied activity within the cells of these plants as the chloroplasts stream around; chloroplasts are the organelles that perform the photosynthesis, capturing energy, converting it and storing it. The number of chloroplasts in a cell varies from plant to plant: a single one in species of single-celled algae, 100 in wheat and cabbage.

So as the process of photosynthesis takes place in this perfectly clear stream at noon, when the sun is at its highest and the light is at its most intense, the essential process of plant life takes place with an exaggerated force. And, by living in water, this is a force that we can actually see with our eyes, unaided by microscopes, with no need to take anything on trust. In a demonstration as good and as magical as anything that David Attenborough

ever showed us – and that's saying something – these plants perform their own trick of magicianship. They fizz. They pass out their oxygen into the water and the water fizzes with what is for us the stuff of life.

The fizzing plants, the water bubbling with oxygen, the process of photosynthesis made visible: I have to say that I found this the most vivid sequence in the marvellous entirety of the filmed *The Green Planet*. Lots of stuff *sounds* more exciting, perhaps especially the calculating and ever-hungry Venus flytrap. But this simple demonstration of the basic process of life, the one process that makes all other life possible, seemed to me a thing of serious beauty and profound meaning. In this stream, life fizzes like champagne.

ABOVE AND OVERLEAF
The secret of life made visible: plants producing oxygen as part of the process of photosynthesis on a tributary of the São Benedito River, Pará state, Brazil.

It is, if you like, a perfect image of photosynthesis... yet we should be wary of the idea of perfection in nature. We like to consider the perfection of a cheetah's run, of an oak in leaf or of a rose in bloom, but perfection is not the point. Evolution isn't seeking perfection: it's about the hope of finding enough for this day and the next: enough to see an individual through to the goal of becoming an ancestor. Evolution will do this any old how. It's not a quest to create a masterpiece. It's more like cooking when you are snowed in:

you must make a series of meals with whatever you happen to have in the larder. Sometimes you might put together a dish of startling brilliance; at other times it might be something that is at least nourishing enough to feed the family. Evolution does not start from scratch with the aim of finding perfection: we have efficient lungs because the first vertebrates to venture onto dry land had swim bladders which had evolved to give them control underwater – but these turned out to be useful when it came to breathing in

air rather than water. The foreleg of a horse, the wing of a bat, the ventral fin of a humpback whale, the forefoot of a lion and the hand that writes these words are all based on the same template.

So it's not about perfection. Perfection is something to associate with memories of classroom geometry, with its perfect circles and equilateral triangles, or perhaps it lies somewhere in the impregnable world of pure mathematics. When we come across something in nature that looks like perfection, we tend to be struck by its incongruity. When we see something that appears to be a perfectly spherical plant, we are entitled to wonder.

Most plants – when they don't exist in the form of seeds – are anchored in one spot. That is how they operate, but it's a system that can leave them

ABOVE
A serious size: measuring a large algal marimo ball in Lake Akan, Japan.

vulnerable: in water, as we have already seen, the levels can change, submerging them or leaving them high and dry. But the marimos of Lake Akan in Japan have a way of moving in response to the changing conditions of the water they inhabit. Technically they are not plants but photosynthesising green algae: they still need the energy of the sun to make food. They are popular with people who keep aquaria, and are known to English-speakers as Japanese moss balls. These marimos can be found in winter all along the icy shore of Lake Akan as flattened green lumps that don't seem terribly exciting to anyone but an algal specialist.

But when the ice melts, they are washed into the water. This can be a dangerous place: the lake attracts flocks of whooper swans. The birds feed by stretching their long necks below the surface; a densely packed wad of algae is highly acceptable. Enough of these wads roll into deeper parts of the lake, out of the swans' reach.

As they roll and shift and collide with each other, so they knock off bits of themselves; it's a speedier version of the process of constant movement that creates smooth round pebbles on a beach. Here, with much softer items colliding, there is a tendency to make things that are very close to being perfect, if rather fuzzy spheres: perfect round balls of life. The successful ones are those that find a safe place: far enough from the surface to be safe from predators but close enough to get sufficient light to photosynthesise. Many of them can exploit the same safe place at the same time: you can find 10,000 gathered together in a cluster; and a total of 600 million has been estimated for Lake Akan. The sight has been compared to a contemporary art installation: certainly it doesn't look like the product of the natural world. The biggest of these balls can be 200 years old.

Being balls, they roll, and that is an advantage. The wind moves the waters of the lake more or less constantly, creating currents, and the currents move the marimos. They also rise to the surface during the day and sink back to the lake floor at night. This is partly because of photosynthesis: the oxygen produced by the process is trapped in bubbles among the filaments, and that makes the balls float. They also vibrate, spin, turn, like something under a grill; this process exposes every aspect of their perfect roundness to the sun, and so they continue to photosynthesise. Other species of algae settle on the marimos but, because they are constantly in collision with each other, jostling as they

respond to the wind-created currents, the competing algae is always getting scraped off. If they had a chance, these invading algae would cover the marimos and block off the sunlight, thereby killing it. But because the balls are always in motion, the invaders don't have a chance.

The phenomenon of the marimos was once widespread. Colonies could be found in North America and Northern Europe, but no longer. They need shallow lakes with uninterrupted wind if they are to survive. These requirements are not too exacting; there are plenty of such lakes in the world. But they also need unpolluted water, and that's not so easy to find in the modern world. A century back, there were many marimo lakes across the

ABOVE
Unnatural-looking nature: water currents shift these clumps of algae about, knocking off the rough edges and creating marimo balls. Lake Akan, Japan.

world. Now there are very few left to show us an eerie perfection that seems almost unnatural.

But when you explore the natural world, that sense of unreality becomes familiar: all naturalists experience a routine difficulty in believing something that's quite clearly happening before their eyes. There's a classic example of this in the champagne stream of Brazil, the waterway that showed us the fizzing, life-giving force of photosynthesis so vividly a few pages back. In these swift-flowing and beautifully clear waters the plants need to anchor themselves to the bottom. Some attach themselves to bedrock, others to individual rocks. The water level varies from season to season and from year

to year, and in the times of plenty the water rises sharply. Plants that were trailing along the surface now find themselves submerged, the water often a metre or more deeper than it was. But the stems and leaves, nourished by all the sunlight available in the crystal stream, are often thick and luxuriant. As a result, the plants are surprisingly buoyant, reaching up towards the surface. A strange thing then happens to those that have anchored themselves to rocks: they stretch towards the surface and as a result of their luxuriance, they float on upwards, carrying their anchoring rock with them. They do this in numbers, so that the sight beneath the surface looks like a balloon race: a series of graceful, bulbous shapes, each carrying beneath them a small, suspended burden.

This is not a miraculous adaptation for survival: this gorgeous time of flight is in fact a disaster for the plants. They drift together, forming clumpy islands, and they die... and as they do so they recycle nutrients into their home

ABOVE AND OPPOSITE
Rock carriers: plants from the pipewort family lift rocks from the riverbed
in a tributary of the São Benedito River, Brazil.

waters, for in nature no death is wasted. These floating mats of rotting vegetation bring down butterflies in hundreds; they drop in like a storm of confetti to suck up nutrients from the lost plants.

One of the drawbacks of having water as your home is that it's not always there. In some places a watery habitat comes and goes with the seasons; ponds, lagoons and small lakes can be full of water and life at one time, and then dry as if they had never been wet. Such lost ponds can be dry for extended periods, and then there is some alteration. Perhaps the river changes course or a long drought is relieved by exceptional rain – and then everything changes: the place is submerged once again. There is an obvious challenge for any plant that makes a home in places where the water comes and goes.

The lotus comprises two species of flowering water plants, and both are sacred to many of the religions of Asia: you can find them in the art of Hinduism, Buddhism, Sikhism and Jainism. They grow – and grow superbly, with complex flowers that can be a foot across – on flood plains, slow-moving

rivers and deltas, and they can be found from India across East Asia; they have also spread, probably by human hands, into New Guinea and Australia. These plants are accustomed to the capriciousness of their environments, and they have a strategy to cope with the fact that water can vanish, and then return again after an extended interval.

Lotuses survive drought in the form of seeds. They drop particularly tough and waterproof seeds, and these can tolerate extended periods of absolutely nothing: a century is nothing to the seed of a lotus. The record is 1,300 years old; a seed dated to that time was deliberately brought back to life.

Other water plants follow a less extreme but still highly demanding form of existence, exploiting their abilities to lie dormant during drought and to sprout again when the water returns. When this happens every year, the land operates on a rhythm that can be predicted, alternating wet and dry. The Pantanal of South America is the world's largest inland wetland, and it covers an area generally estimated at around 70,000 square miles, about twice the size of Portugal. Several rivers flow into this vast wet sponge of a place, ending

ABOVE
Delicate meal: northern jacana feeds from a lily;
Phinda Game Reserve, South Africa.

OVERLEAF
World of water: seasonally flooded fields in
the Pantanal, southwest Brazil.

OPPOSITE
Beautiful error: a waterlily has opened underwater
and so is unavailable to pollinators; Ain, France.

their journey not in the sea but in this great green endlessness: rivers that include the Taquari, the Miranda, the Negro and the Cuiaba. These rivers swell massively in the times of the rains, and transport an immense volume of water into the Pantanal, which, as the water rises, becomes annually the largest flooded grassland in the world.

It is a place of immense productivity, immense bioabundance, immense biodiversity. More than 9,000 species of invertebrates have been found here, and the vertebrates include around 450 species of birds, around 250 each of mammals and fish, and getting on for 150 species of reptiles and amphibians. As always, this vast collection of animal life wouldn't exist without the plants. As the waters rise every year, so the water plants must take advantage: those that win the race for space have earned the right to live and the chance of becoming an ancestor. The water is rich in minerals and nutrients, dark with sediment. It offers many opportunities for life, those that can seize them are winners – and it's tempting to burden them with adjectives like aggressive, pugnacious even vicious.

The surface is the place to be. The light can't penetrate very deep in these murky waters, so you need to be on the top if you want to photosynthesise. The water will only last for a few months in these generous quantities, so the water plants must live their lives in a hurry. If you take your time about things, your habitat will have vanished before you have fulfilled your biological destiny. And if you are meek and gentle in your growing habits, you won't be able to survive the rush to fill the available space and make more of your own kind. The plants have developed different strategies and structures: buoyancy chambers, floating leaves, bladders on roots, all devices that will shoulder aside neighbours so their owners can fill the newly created gaps with their own leaves. It's a fight for territory, which for an unmoving plant is the same as the right to exist. Some plants get smothered, others get pushed into places of opportunity, finding new channels, or drifting downstream, sometimes to find new opportunities, sometimes to fail.

Water lettuce is a familiar plant to many of us. It has, with human help, invaded many waterways across the warmer parts of the world, where it has become a problem, covering entire water surfaces and blocking light from the water it sits on. In the Pantanal, with its seasonal rises and falls of the water, it is a native and functioning part of the ecosystem. It grows on the surface in a

OPPOSITE
Little room for competitors: giant water lily, Pantanal,
southwest Brazil.

nice little green rosette, quite a lot like real lettuce, though it's also called Nile cabbage. But, in the manner of an iceberg, its most important characteristic lies out of sight, below the surface of the water. Each plant has a massive tangle of roots hanging from the plant, not reaching to the bottom of the waterway but extracting what it needs from the water itself. These roots seem disproportionate to the plant's needs, altogether too much of a good thing. But the roots do more than nourish the plant to which they belong: they also deny space to any other plant attempting to come near. They seize most of an entire cylinder of water beneath the water as the plant's personal space: a place where nothing can grow except itself, so that the leaves on the surface can develop and the plants can eventually set seed.

ABOVE
Food and shelter: water hyacinth in its natural environment,
Amolar mountain range over the Pantanal, Brazil.

They also propagate vegetatively: by cloning themselves. A spreading colony of water lettuce produces a thick entanglement of roots below the surface, and that inevitably provides opportunities for other species. Water lettuce roots become a nursery for the offspring of many of the species of piranha that live in the Pantanal. I should add here that most of our beloved legends about piranhas are not true: they don't lurk around South American rivers in vast gangs waiting to strip cows – and humans – to the bone in seconds flat. They are carnivorous, sure, but they mostly swim in shoals to find safety in numbers and protect themselves from their own predators. They also show strong parental behaviour: rather than broadcasting millions of eggs and letting them fend for themselves, as many fish species do, piranhas lay their

eggs in the roots of water lettuce and both parents will guard them until they have hatched.

The water hyacinth grows alongside the water lettuce, competing for the same space and operating a similar strategy: grow fast and cover the surface of the water, thereby taking control. Like water lettuce, the water hyacinth has been introduced all over the world and has become a problem: the plants naturally seek to dominate any ecosystem they find themselves in.

But in the Pantanal they are part of the natural and annual course of events and, being plants, they provide nourishment for many other species. During the watery times of the year, when life is at its most abundant, unlikely creatures take advantage of what is on offer. In the southern Amazon, forest monkeys come down from the canopies to savour the annual bonanza of the waterways. There are many dangers below the surface, including caimans and anacondas, so the monkeys keep well clear of the water while still finding the opportunity to feast on its largesse. Spider monkeys got their name because of the strange illusion that comes from the way they move: they use their prehensile tails as easily as any of their four limbs, and so they can look like giant multi-legged spiders as they move through the trees. It's perfectly possible for spider monkeys to hang from their tails alone and, with both hands free, collect water hyacinths from the surface. With their clever fingers they can dissect the plants and eat the best part: the fleshy heart, and for additional protein, the invertebrates these usually contain.

The hoatzin is a bird that also devours these leaves, and is as bizarre a bird as has ever hatched. There was a sequence starring the hoatzin in David Attenborough's first great blockbuster series, *Life on Earth*, which was first aired in 1979. It told the story of the evolution of life by using the animals that still survive. *Archaeopteryx*, the earliest bird to have been found as a fossil, had claws on its wings; so too do young hoatzin. Attenborough brought us footage of these strange birds climbing around the branches like monkeys.

Their way of eating is as peculiar as the furnishing of the wings of their offspring. They eat leaves, and have a complex multi-chambered digestive system that allows vegetation to ferment inside them, a little in the manner of a cow. A diet like this requires bulk feeding, and birds that fly tend to avoid bulk because of the problem of excess baggage that affect every flying device. A

OPPOSITE
Living fast: water hyacinth in the Pantanal, southwest Brazil.

hoatzin's stomach can add up to a quarter of the bird's total weight, and it must sit still and allow its gut flora to get on with the work of digestion while it rests its belly on a convenient branch. This is a process that inevitably creates a lot of gas: the birds have been nicknamed flying cows – and also stink-birds.

The hoatzin operates on a design unique among birds and that's what allows the species to compete for a place in the world, and to hold on to it and prosper. Competition need not be overt and aggressive: acting like a feathered cow and resting your belly on a branch between meals is not exactly warlike, but it is a strategy that allows the hoatzin to make a living.

Other species adopt a more belligerent approach to the problems of life. That can be as true of plants as of animals, as we have already seen – but no plant demonstrates the principle more dramatically than the giant waterlily: the giant whose enormous floating leaves seem always to be photographed with small children sitting on them, as safe from the water as if they were in their mother's arms.

Readers with long memories may recall *Watch with Mother*: programmes for very young children that were shown on the BBC. Each programme began with an opening flower, a gentle anticipation of the sweetness that was to follow. Time-lapse photography has allowed us many insights into the lives of plants, and the miracle of the opening flower is a classic example. Such memories contrast sharply with a sequence in *The Green Planet* that shows the development of the giant waterlily: a serious contender for the most dramatic footage in the entire television series.

It begins with the appearance from the water of a spiny bud that resembles a spiked mediaeval mace or, to be more technical, a morning star, a weapon that was wielded from a long handle with devastating effect. The vegetable mace is just as effective: the time-lapse reveals a series of threshing, circular motions of the mace that clear the water of competing plants. The leaves then grow into the cleared space, and they are gigantic: often more than two metres across. They have spines on their lower surface to protect them from fish and others that might be inclined to eat them. The great leaves roll over the top of competitors, blocking off the light, or they push other plants aside.

The leaves are a marvel of design. Joseph Paxton borrowed the structure of the leaves of the giant waterlily to create his masterpiece: the Crystal

OPPOSITE
A process of domination: a giant waterlily in bud (top left), a lily pad unfurling (top right), fully opened (middle) and in full flower (bottom).

Palace, the immense building – four times the size of St Peter's in Rome – that held the Great Exhibition of 1851. The leaves stretch across the waters of lagoons, covering the surface, hogging all the light and using those immense leaves for photosynthesis. They can grow up to half a metre in width in a single day; a single plant can have a good 40 leaves. The long, extendable stalks keep the plant precisely where it needs to be: on the surface, whether the water levels rise or fall. The leaves have a raised centre so that any water that lands on them gets drained off quickly; a layer of water would compromise the efficiency of their photosynthesis. Any water that's left is drained out through cylindrical pores called stomatodes.

This is a marvellous plant that dominates a lagoon as an emergent rainforest giant dominates the area all around it. The waterlilies will take on others of their own kind, rolling their leaves over those of their rivals, forever seeking light. Plants that live on the surface of the water can only expand in

ABOVE
David Attenborough discusses water crowfoot on the River Avon, Wiltshire, UK.

OPPOSITE
Water plants like water crowfoot must get their flowers out into the open air if they are to be pollinated; Netherlands.

OVERLEAF
It's about timing: red lotuses flowering together, overflown by an egret; Nong Han, Udon Thani, Thailand.

two dimensions, unlike the three-dimensional possibilities of those rainforest giants.

The Amazonian waterlily has always fired the human imagination. They can be viewed in England at Kew Gardens: a house devoted to this single species was opened in 1850. I remember childhood visits to Kew when this, the house of mystery, seemed always locked, the plant not in flower so unworthy of visitors. I remember, too, the long-awaited day when I could enter and gaze on this impossible plant. Alas, it was not permitted to see if one of the leaves was up to my weight.

Water plants can find problems when they attempt to make more plants. It is relatively straightforward to reproduce by cloning; to do so sexually – to produce flowers and to get pollinated – can be more complex. Water crowfoot, which grows in Europe, the west of North America and in Northwest Africa, solves the problems of shifting water levels by its floppiness: the stems bend easily, and so they stay in contact with the water, mostly under the surface. That's fine for a stem, so long as the water is clear

enough to permit photosynthesis, but it's no good for a flower that needs to be pollinated by airborne insects. So the plant changes its pattern, and grows its flowers on rigid stems. These rise above the level of the water and attract the insects they need to propagate themselves.

There is an optimum time for flowering in most environments, as we have already seen in deserts and in seasonal lands. A water plant needs to maximise the best water conditions that are suitable for its own purposes, at the time when there are the most pollinators available. Ideal timing is no more a secret in water than it is in other environments. Watery environments produce their own spectacles of annual bloom, in which the plants are both competing with each other for pollinators, and at the same time cooperating with each other to bring as many pollinating animals to work as possible. In some places, flowers follow the receding edges of streams, ponds and lagoons as the waters begin to shrink, creating meadows of flowers that seem to be pushing the annual floodwaters back, doing so in a great hurry before the grasses that dominate in the drier parts of the year can re-establish themselves.

What is the most spectacular annual bloom of them all? It's one of those personal choices, and this book offers a fair old sample to choose from – but certainly one of the stronger contenders is the massed bloom of lotuses. These plants can dominate waterscapes with thousands of simultaneous blooms. They are also capable of doing something still more remarkable, if not quite as spectacular. The oxygen needed by plants can be hard to find in many stretches of water, for the concentration is much lower than it is in open air. But the waters in which you find lotuses are very short of oxygen indeed, and the mud in which you find the rhizomes are almost completely anoxic. But it has solved the problem by creating an extraordinary system of gas canals, in which oxygen is taken down from the air by older leaves on the surface of the water, to be transported down to the rhizomes in the mud below. The air is then passed out by other leaves, a system that operates like air conditioning.

Water plants must sometimes take extravagant measures to make sure they are pollinated. The giant waterlily has also come up with a dramatic advance in plant technology to get the best out of its pollinators: it produces flowers that are, in effect, hot-blooded. Among animals, mammals and birds generate their own heat; most of the rest acquire their heat from the warmth of the sun. Making your own heat is expensive in terms of resources – the lifestyle needs a lot of fuel, which we take in as food – but it offers warm-blooded creatures a versatility that the rest of life can't rival. We have seen, in the previous chapter, the way the daisy follows the sun, in order to present a pleasantly sun-warmed flower to the world, one that makes it more agreeable for a pollinating insect to land on. The giant waterlily takes this a stage further by generating its own heat.

The plant produces flowers as magnificent as its child-supporting leaves: flowers as big as footballs. They only last two days, so they need to be extremely efficient at attracting pollinators, and they also require the pollinators to be as efficient as possible. The plant achieves this by first attracting its pollinators and then by holding them captive while they perform the task of pollination.

Thermal imaging shows the flowers are warmer than the surrounding air by 10°C. This alone makes the flowers desirable places of refuge, but the warmth also pushes out the scent of the flowers, creating a very powerful double inducement for insect visitors. The twin strategies attract beetles, which fly in and get down at once to the warmest part of the flower, at its base, where they can also feed on the plant. On the way down, they pass the stigma or female part of the flower, which receive any pollen that the beetles may be carrying from another lily. As the beetles reach the bottom of the flower, the flower closes, trapping them in there for the day. This is no great hardship for the beetles. Quite the reverse: it is a matchless opportunity for them to feast, fight and fornicate. The beetles feed, compete and mate in the warm sticky closeness of the flower. The following day, the flower heats up again, but only at the tip, drawing the beetles upwards and, at the same time, warming their bodies and energising them so they are in the best possible condition to fly. The flowers open, and the beetles fly off, newly covered in pollen, to find another flower. The flower can hardly do more for its pollinators: feeding them and then directly helping them to fly and seek more

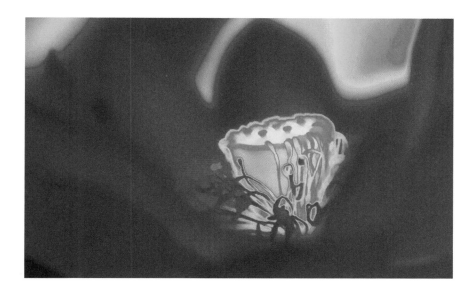

flowers. After that the white lily turns pink and then purple, and dies, its job done, while the seeds develop in the capsules of the late flower. The plant itself survives – and is, with luck, on the way to making more of its own kind.

The fundamental principle of life for most plants is immobility. Plants are constants, literally rooted to the spot while the rest of life carries on all around them. Plants are the fixed points of the living world, coming and going in the course of time, but throughout the course of most of their lives they are committed to a single spot. This, as we have already seen many times over in this book, requires plants to be inventive when it comes to propagating themselves. They need mobility to spread and receive pollen. They borrow mobility from the wind or from convenient animals, which they often bribe heavily with food and drink.

They also need to find mobility when it comes to spreading their seed and, again, both wind and appropriately bribed or tricked animals offer solutions. Plants that live on the banks of watercourses often drop seeds into the water, and the water bears these away. But sometimes a journey downstream, carried towards the river mouth by the inevitable seaward flow of a river, is counterproductive, taking the seeds to places where its future is

ABOVE
Heated flowers: an infrared thermographic camera reveals the self-generated warmth of the sacred lotus.

compromised by a less sympathetic environment. In some cases, the best thing for the plant is to float its seeds upstream, against the flow and towards the source. This is not the flat impossibility it seems, because plants associated with water can recruit animals unavailable to those that live in dry places. Some plants propagate their seeds by exploiting fish.

A species of fish, known locally around Bonito in Brazil as piraputanga, are committed to a migratory lifestyle. They come in numbers, swimming upstream as the floodwaters rise. Unusually for fish, they are great consumers of fruit. They follow troupes of monkeys feeding on the trees that line the waterways; the monkeys pick and eat fruit and what they let fall is food for the fish. This passive method of gathering fruit is effective enough, but the piraputanga can also collect fruit unilaterally. The idea of a fruit-gathering fish seems a perfect contradiction, but it's relatively common, especially in South America. And it's a way of life for the piraputanga: they leap clear of the water to pluck fruit from branches that hang over the river. This requires a mixture of power and accuracy. The leaping fish must calculate distance, and to do so it must deal with the way that light behaves differently in water and in air. The fish must close its jaws at the precise moment it comes in contact with the fruit. It's quite a trick and one that works to the benefit of the plants as well as the fish. It's as if the plants were exploiting the fish to go to such extreme lengths to disperse their seeds, but the fish, bribed with the rich flesh of the fruit that wraps the seed, are also onto a good thing.

The seeds, consumed with the fruit, take a few days to travel through the digestive system of the fish. By the time the seed is passed out at the other end, the fish is likely to be several miles away from the tree that bore it – several miles upstream. The fish has obligingly battled the river's current, gone against the flow and, if all has gone well, deposited the seed in a place that suits it far better than a random berth in the opposite direction.

Making a fruit is a fairly serious commitment. That's because it takes a good deal of energy and resources. Nature often seems like the cynic beneath the Christmas tree, calculating the net worth of every present. It doesn't matter how thoughtful, well-chosen and morally generous the gift is: the only thing that matters is the price. Evolution, acting over uncountable millennia, provides the most searching and rigorous cost-benefit analysis that could ever be carried out and, in this case, the monkeys and fish won't perform the

OPPOSITE
How to get your seeds upstream? Piraputanga fish do the job at Bonito, Brazil: eating the fruit and passing out the seed as they travel against the current.

OVERLEAF
Greater reedmace and common reed in a watery landscape; Somerset Levels, UK. Glastonbury Tor is visible in the background.

services the plants need unless they get paid. But some plants take an opposite strategy and invest not in quality but in quantity. Bulrushes, sometimes called cattails, do just that. Their seeds get well dispersed, even though they don't pay any dispersers to do the job. Instead they use the wind. The wind will do what it does, regardless of reward.

The advantages of paying a monkey or a fish are clear enough: either of these consumers of seeds is likely to place a seed in the right place, one that will give the seed every chance of germinating and growing. The monkeys pass seeds out onto the forest floor; the fish take seeds upstream and deposit them in the water, where they have every chance of finding the shore and germinating. This is a precise and targeted strategy: the trees produce comparatively few seeds, but there is every chance that a seed-dispersing animal will plant one in just the right place. It is a strategy of narrowcasting.

Bulrushes go in for broadcasting. The attractive brown heads of the rushes – they are sometimes also called reed mace, so that's the second plant to be compared to a mediaeval mace in a single chapter – are packed with seeds. A single seedhead can contain 220,000 of them, which is a little like placing a bet on every single number on a roulette table. A sequence in the televised version of *The Green Planet* shows us Attenborough in action again. He took the stout, velvety seedhead of a bulrush in his hand and gave it a good old twist, releasing countless seeds into the air, to land wherever the wind might take them. When given a tap, the seedheads will give a soft explosion, sending more clusters of seeds off to seek their fortune.

While simple luck is a good part of the plan, the bulrushes have a further strategy to twist the odds in their favour. Rushes grow with their feet in water, and it's obvious that the wind will dump many seeds on the water. If the

bulrushes are growing around the edge of a decent-sized lake, a good many seeds will land on the water. You can make as many seeds as you like, but if they all drown, you haven't furthered your cause very far.

The bulrushes have an answer to this: they make little sailing ships of every seed that lands on the water. At first the seeds all sink, which looks like a disaster for the plant. But not so, because the submerged seeds grow threads of mucus and, with this added buoyancy, they rise up to the surface again. Once there, the mucus acts as a sail, and the seeds are propelled across the surface of water, moving until they can move no more – that is to say, until they reach the edge. There they will sink once again, like as not in shallow water surrounded by their own kind, in the right environment for a bulrush to grow and prosper and produce seed heads of its own, each one capable of releasing close to a quarter of a million seeds into the eternally willing wind.

But for a plant that is totally committed to water – rather than living on the margins – the process of flowering and seeding is difficult and expensive. As we have already seen, the making of flowers and fruits requires masses of energy and nutrients, and these are often in short supply in the watery life. To

make things a great deal more difficult, pollen – the male substance in plant sex – doesn't usually work when it's wet. That simple fact makes it seem that a water plant is doomed from the start. But that is obviously not the case, even though many water plants flower and seed only rarely. For them, sex is not a crucial driving force: it is more of a back-up.

So let us look at duckweed, the plant that grows on the surface of still and slow-moving water, often covering it so completely that a stretch of open water looks more like a freshly mown lawn. You'd think you could walk across it in comfort; when you row a boat through it, you part thousands of tiny plants, leaving a broad trail of clear water, which you can see closing up behind you as you row on. Duckweed can indeed produce flowers and seeds but, far more often, it goes in for cloning. It makes copies of itself. We have seen the creosote bush of Mojave Desert cloning itself, making rings of

OPPOSITE
How many seeds in a single bulrush? David Attenborough releases about 20,000 of them.

ABOVE
A secondary use: a yellow warbler collects seeds from bulrush cattail to use as nesting material; Montana, USA.

bushes around itself from what is essentially the same plant, doing so in slow motion so that the process is still continuing 7,000 years after the plant was established. Duckweed has the same idea, but it does it rather quicker.

A single duckweed plant can make 17,500 copies of itself in two weeks, an alarmingly precise figure. Exponential increase is a mind-boggling phenomenon: further calculations say that if left unchecked – an impossible proviso, of course – a single plant could cover the entire surface of the planet in four months. A galloping increase, a visual lesson in advanced multiplication, as from humble beginnings the plant sets out on a maniacal all-conquering spread.

You wonder, then, why duckweed doesn't simply carry on and take over the real world. One reason is that the plant is fairly precise about its habitat needs: it must have slow or still water in fertile conditions. Another is that the plant is much eaten: it is a high-protein staple for many species of waterfowl, that is to say ducks, geese and swans. For them, it is an excellent resource: duckweed holds more protein than soy beans. The cover it offers so generously across the water is an advantage to many other creatures: it hides fry, the very young fish, and offers shelter to mature fish and frogs, and it reduces the evaporation from the water surfaces it covers. It also suppresses blanket weed that would otherwise choke a pond.

The nature and being of this plant is a perfect example of enlightened simplicity: its ancestors were far more complex but, as time progressed, these ancestors found increasing advantage in added simplicity rather than increased complexity, much as we have already seen in these pages with the parasitic plants like dodder and desert mistletoe. Duckweed is, as it were, a wilfully simple plant, one that has had the complexities refined from it in the course of evolution. One of these simplifications is the way it mostly propagates itself by cloning. This is a natural version of the gardener's technique of propagating plants with cuttings, by taking a piece from one plant and making it grow into another plant, thus producing two quite separate individuals which are at the same time genetically identical.

The principles of vegetative reproduction, and its vivid success in the example of the world-filling duckweed, ask us intriguing questions about the way that life works. If such a simple system – never mind the complexities of sex, just make more individuals identical to yourself – is so successful for the duckweed, why don't all plants do it? Why don't we animals do it as well? The answer lies in the vulnerability of monoculture, an issue we will look at again in the next chapter. A vast colony of genetically identical individuals has no resilience. It has no resources when disaster strikes. If a disease hits the colony, there will be no outliers immune to the disease: if one goes, all do. If conditions change – if, for example, the environment gets measurably hotter – there will be no individuals better suited to that change: they will all be equally disadvantaged, all in the same boat. You have avoided the energy-expensive complexities of sex, but you have compromised yourself by doing so. There is always a payback, and that is why the planet-conquering duckweed is also a fragile thing after all.

The concept of universal fragility is a new thing in the history of human thought. Animals we thought terrifyingly powerful have been brought to the brink of extinction, plant systems like rainforests that seemed endless and eternal have been reduced to patches and pockets, and the entire planet, whose resources once seemed capable of nourishing its dominant species for ever, is

OPPOSITE
Deforestation upstream has caused erosion; the resulting silt
clouds the Betsiboka River, Madagascar.

suffering the traumas of climate change. The idea that humans could destroy the global ecosystem on which we depend is shockingly new: most people prefer to look away from it.

Water plants are by nature fragile. It's been that way since they first began to evolve hundreds of millions of years back. Most of them require very specific if not unique conditions if they are to thrive: the duckweed we have been discussing is not much good if the water it floats on starts to move beyond a dawdle, or if the pools or watercourses dry up. Many plants that thrive in rivers upstream are unable to make a living where the river advances towards the sea. The channels become too deep for them, the waterflow is more powerful and suspended silt blacks out the light, making photosynthesis impossible. Where the river changes its nature, many plants are left behind. That's the way things are.

These days, humans are constantly changing the nature of rivers and everything else on the planet, freshwater ecosystems perhaps more than anything else. Reckless deforestation upstream means that water is no longer held up when it falls from the sky. Instead it rushes into the river bringing sediment with it, darkening streams that were once clear. Dams change the

seasonal flow of the river, dramatically altering ecosystems both above and below the dam, changing the unique geology and chemistry of the river itself. It's been claimed that dams cause more plant extinctions than deforestation: in the United States there are 9,265 dams, in China 23,842. Their construction creates greenhouse gases, and so does their maintenance. They destroy carbon sinks, deprive ecosystems of nutrients and destroy habitats. They were built with good intentions but, once again, we are left with the law of unintended consequences.

We humans have the capacity to perform miracles and we do so on a daily basis. In many places we have managed to turn rivers into deserts, at least so far as flowering plants are concerned. But as the rivers slow when they reach the end of their journeys, there is another opportunity for plant

ABOVE
Polluted tributary to the Li River, Guangxi, China. Town and mountains are visible on the horizon.

growth. Where the rivers meet the sea, they lay down sediment and that provides an opportunity for seagrass: a family of flowering plants that adapted to a marine environment about 100 million years ago. Seagrass thrives where there is good sediment and plenty of sunlight to permit photosynthesis. The plant rises from rhizomes that grow just beneath the seabed. This gives it an unexpected advantage over seaweed in sandy places: the rhizomes, which evolved for life on land, can allow the seagrass to get a purchase on the seabed where seaweed can't. The plants can spread vegetatively by pushing out fresh sprouts from these rhizomes, but they can also reproduce sexually. Unlike the water plants we have met so far in this chapter, they flower underwater, doing so at night to avoid being eaten by parrot fish. Here they release pollen – so they are exceptions to the rule that

ABOVE
Underwater fields: covering about 4,000 square kilometres of seabed, the largest seagrass meadows in the world are at Shark Bay, Western Australia.

ABOVE
Semi-vegetarian shark: bonnethead sharks feed on seagrass;
Florida Keys, USA.

states that pollen doesn't work when it's wet. The pollen is spread by the ocean currents, and also by small crustaceans that have been given the name sea bees.

Seagrass spreads out to create expansive underwater meadows, and they are as rich a food source as any grassland system on dry land, perhaps even better. The largest of the animals that feed here are the manatees and dugongs, four species of large, slow-moving aquatic herbivores that graze these meadows, cropping the grasses with their split upper lips. Readers with long memories will recall David Attenborough's *The Life of Mammals*, first broadcast in 2002, in which Attenborough, standing in shallow coastal water surrounded by manatees, visibly quails at their aromatic breath. They can eat 50 kilos of seagrass in a single day, more than 10 per cent of their considerable bodyweight. The seagrass meadows are also grazed by turtles and by a species of semi-vegetarian shark, the bonnethead, which has a diet that is 60 per cent seagrass. It's been estimated that a single acre of seagrass can contain 40,000 fish and 50 million invertebrates.

The seagrass plains are important habitats for all of us. They bind and stabilise the seabed, they store carbon, stopping it from being released into the atmosphere as one of the principal greenhouse gases (one calculation estimates that seagrass across the world sequesters enough carbon to offset all the vehicle emission in the UK every year). Seagrass covers 0.1 per cent of the ocean floor; the plains hold 11 per cent of the organic carbon found in the ocean.

It is the wearying duty of every writer on natural history to follow a presentation of wonders with a few sobering moments of gloom. Few of us do so with relish: it is a sad duty, and one it would be irresponsible to shirk. So once again I must note that this important habitat is,

OVERLEAF
Rare sight: an exceptionally large herd of dugongs grazes seagrass in Moreton Bay, Queensland, eastern Australia.

yes, under threat and diminishing as a result of human activities: disturbance, pollution, and overfishing. Here's another glorious calculation: in the 1500s, the seagrass areas of the world are thought to have held 20 times more large grazing animals than the Masai Mara.

It doesn't take a scientific genius to work out that freshwater environments are important for human beings. In recent years, western civilisations have made a fetish of fresh water: many people are unwilling to take a step away from home without a plastic bottle containing fresh water, the bottle to be thrown away when the water has been drunk. The more recent trend for reusable containers still emphasises the point that we humans can't survive without fresh water. The same thing is true of every other species of animal that lives outside the oceans. Water plus sun equals life: we encountered that basic fact right at the beginning of this book. It follows that the most precious things in our lives are the sun that stands 148 million kilometres away from us and the fresh water that we can find all over the land areas of the planet we inhabit. The sun is safe for the time being, though we have found a complex way of allowing it to increase the effects of its heat on this planet, to our own increasing inconvenience. Still, the sun itself is beyond our range when it comes to active harm. Alas, fresh water is not.

In this chapter, we have been celebrating some of the great botanical spectacles of the world, from the plants that inhabit the tumultuous streams of the high mountains to those that grow where land and ocean meet. In the colour – and especially the colour green, the colour of photosynthesis – we have been able to rejoice in the plants that make fresh water their home. Again and again, they reinforce the truth that fresh water and life are inseparable things, so far as we terrestrial animals are concerned.

Fresh water is of immense importance to humanity: our lives would be impossible without constant daily access to the stuff. So perhaps we need to recruit the five-year-old child who invariably sees things more plainly than the grownups, and listen to the question this child must ask: if fresh water is so important to us, why are we destroying the places where it is found? Why is it that fresh water is one of the most threatened environments on Earth today? (And the competition for that particular accolade could hardly be more intense.)

OPPOSITE TOP
A bridge of boats. Water hyacinths have blocked the Buriganga River and hampered the movement of boats in Dhaka; commuters travel to work by this temporary bridge

BOTTOM *Propeller scars in a bed of seagrass, resulting in long-term damage; Florida Keys, USA.*

Pollution, dams, deforestation, increasing sediment – all these things are not only leading to the extinction of plant species, they are also destroying the system that allows water plants to take in carbon dioxide and pass out oxygen into the atmosphere, as well as storing carbon. We recklessly take water from the earth, causing the fresh-water systems to dry out. We deliberately drain wetlands for agriculture and other developments. We allow water systems to be invaded by exotic plants. The frightening acceleration of global heating has led to increasingly frequent episodes of drought and flood. In this way, we make an enemy of what is most precious to us.

It all comes down to the eternal human propensity for short-term thinking: whatever it is that gets you through the day must be all right. Politicians in democratic countries think only about the next electoral term; those with more autocratic government systems do whatever they believe will keep them in power. Why should they bother about the future of the human race? Why should they trouble their heads with thoughts of the world their great-great grandchildren will inherit? It seems that we manage the fresh-water systems of our planet on the guidance given in a grim and ancient joke. What has posterity ever done for us?

RIGHT
Where turtles may safely graze: green sea turtle feeding on seagrass; in the background a researcher for the Centre for Ocean Research and Education; Bahamas.

HUMAN
WORLDS

The history of humanity is the story of the relationship between plants and people. It was a vital relationship for the first humans who walked upright on the savannahs of Africa; it's equally vital for the most dazzling twenty-first-century urbanite. Plants bring us all of our food: when we're not eating plants, we're feeding from creatures that eat plants. No animal can be an animal and no fungus can be a fungus without plants. Plants also bring us the air we breathe and the water we drink. This is the Age of Plants, and we humans will never move beyond it.

We have altered the planet we live on beyond recognition so that the plants we choose can grow in the places we choose to grow them. Plants that we don't want have taken advantage of our energy, our clumsiness and our lack of vision and all over the world they grow in places where humans don't want them. You can find Himalayan balsam in the beauty spots of England and Southeast Asia's kudzu all over the American south.

This was always the green planet. Ever since life began, Earth has been the planet that runs on plants. The first photosynthesising organisms – cyanobacteria – have been dated to 3.4 million years ago. In the current century, the whole planet reflects the choices that humans have made, and continue to make. The truth that guided the lives of the first humans still operates today: we travel between birth and death by means of a bridge made from living plants.

That truth becomes easy to understand – blindingly and inescapably obvious – when we travel to the northeast of India, to Meghalaya state and visit the Khasi, a tribal people who live in this precipitous and heavily forested place. Meghalaya means abode of the clouds in Sanskrit, and a good name it is too. The place is full of lofty hills and rains that seldom let up: at least 12,000 millimetres fall there every year, making it the wettest part of India – no small achievement. Meghalaya is marked by steep ravines, lavish waterfalls and sleep slopes in perpetual dangers of mud slips and landslides. Getting from place to place in such country should be ridiculously difficult, but it is not. The going is made easier by a vividly harmonious relationship between humans and trees, and it takes the form of bridges that live.

Many species of fig trees start their lives on the branches of other trees, with roots that grow downwards, taking in moisture from the air as they do so, eventually reaching the ground, if all goes well, to become independent

OPPOSITE
Growing free, by kind permission of humanity: ferns, mosses and blooming azaleas in Sapsucker Woods, New York State, USA.

structures. These roots are tough, versatile and accessible, and for centuries the Khasi people have used them to manage the landscape for human use. The roots are twisted and trained and directed in ways that allow them to hold the vulnerable slopes together against the danger of landslip, and to create ladders that make climbing vertical slopes more straightforward. But above all they make bridges: bridges across ravines, made from the living roots of fig trees. These roots maintain their thickness as they grow through the air, but start to thicken when they hit soil. Their immense tensile strength allows them to take control of their host-tree in normal life – and humans exploit that quality to make bridges.

A sumptuous time-lapse sequence in the televised version of *The Green Planet* shows us how it is done. The roots are trained through bamboo staves split in half to make U-shaped troughs: the roots march along routes pre-ordained from them. The technique depends on finding two fig trees, one on

ABOVE
Crossing-point: a living double-decker bridge made from fig-trees by the Khasi people of Northeast India.

each side of the gorge you need to cross. If there is no obliging tree where you need one, you plant it. The fig trees grow and their roots eventually usurp the temporary scaffolding of bamboo and grow onwards. It takes time; the current bridge-makers are working for generations yet to come. The longest span crossed by such a bridge is a good 50 metres; the oldest perhaps 500 years. These bridges are far more durable than anything you could make from cut wood or modern material in this extreme landscape and extreme climate. These bridges can't rot because they're still growing.

The Khasi people have a myth: there was once free travel between Earth and Heaven until a sacred tree was foolishly cut down. It's a story about living rightly, with a sensible, self-preserving attitude to nature. The living bridges of the Khasi people make a powerful symbol: without living plants we will all fall into the abyss.

The fig trees that make these bridges are classified as flowering plants, even though their flowers are not, for us humans, the most significant thing about them. In fact most of the plants that concern us have flowers. There are about 350,000 species of flowering plants – technically angiosperms, or 90 per cent of all plant species; this includes many plants we don't think of in terms of flowers, including grass species (like wheat and other cereals). Flowering plants are called angiosperms, which means that their seeds come wrapped up in a fruit; gymnosperms come with their seeds unclad: a gymnasium was originally a place where people went naked.

Though flowering plants evolved as gradually as anything else, they underwent a great explosion of range and diversity 100 million years ago. Many of them evolved truly spectacular flowers. They did so to attract attention: attracting attention was nothing less than the new survival mechanism. They were no longer using the wind to spread their genetic material, a system that is unpredictable, uncontrollable and lacks a policy for targeting. Now they were beginning to exploit species from the animal kingdom to do the job, and to do it with some precision. These animals could be guided accurately onto the target by making the target spectacular and unmissable. They did this by coming up with a new kind of leaf: leaves that came in colours that stood out dramatically from the sea of green, busily photosynthesising leaves that surrounded them. These modified leaves are petals, surrounding the plants' reproductive organs and sending out a

powerful message to pollinators: *I'm here! Land here! What you want is right here!*
The petals, the glory of a flower, entice the creature to fixate on the flower,
often to take the proffered bribe – or fair and just payment – of energy-giving
nectar and / or highly nutritious pollen from within. (Though some flowers
cheat, offering false advertising that cons the nectar-seeker into doing the job
for nothing.) It then moves on to another flower, preferably one of the same
species, to deposit some of the genetic material – the pollen – it has gathered,
on purpose or by accident. Under this system, the genetic material is placed
with infinitely more efficiency and purpose. It was a development that led to
an explosion of diversity. Plants initially had small and inconspicuous sex
organs: now they provide the greatest show on Earth, as we have seen again
and again in these pages.

Flowers evolved to please insects and other pollinating animals. But they
also please us. They please us profoundly, so that we use flowers to mark the
most significant parts of our lives: love, marriage, birth, death,
enlightenment, prayer, hopes of an afterlife and the desire for peace. The
bride's bouquet, the wreath of poppies on the Cenotaph, the sheepish bunch
of roses on St Valentine's Day – flowers mark the course of human life, in
different ways in different cultures, but often enough as a gift from a male to
a female. There is something so deep about this that flowers seem more than
beautiful living things. They seem to represent the meaning of life.

Flowers play a central part in many of the most significant days in every
culture on Earth. In Mexico at the end of October they begin the
preparations for the great festival known as the *Dia de los Muertos*: the 'Day of
the Dead'. That involves the harvesting of millions and millions of flowers:
most of them marigolds. Marigolds – the species known as *Tagetes erecta*, or as
Mexican, Aztec or even African marigolds – are native to Mexico, and are
now grown as a commercial crop, most of them for this, the day of days,
when the spirits of the beloved dead return to Earth. It is a day of celebration,
not of mourning, its theme is the continuation rather than the ending of life,
and it all takes place to the background of marigolds, flowers that stand for
the warmth and the light of the sun: for life itself. The marigolds are used to
create avenues that guide a soul back to the family altar; as flowers attract
pollinators, so, here they are used to attract human souls. The sugar skulls, a
traditional part of the Day of the Dead, have a special fascination for people

OPPOSITE
TOP *Best-laid plans: the Indian
paintbrush plant attracts a devourer – a
hoary marmot – instead of an insect
pollinator; Mount Rainier National
Park, Washington State, USA.*

BOTTOM *Pleasing sight: many plants
please insect pollinators to their profit
and often inadvertently please humans
at the same time. Orange lily, meadow
clary and yellow rattle in an alpine
meadow in the Dolomites, Italy.*

from different cultures, but the marigolds are the real heart of the festival. The golden orange colour makes it inescapably clear that this is a celebration of love and life.

When Europeans reached America, it was the beginning of what is called the Columbian Exchange. The marigolds of the Americas travelled to Europe and Asia, along with plants like chilli peppers, potatoes and tomatoes, while wheat, rice, apples and oranges went the other way. Marigolds are now central to Indian religious life. They have only been there about 350 years, but it's as if they have always been part of the most ancient religious rites. In India, almost everything is given a little added holiness when draped with a wreath of marigolds, woven onto a thread: a shrine to the Lord Ganesh, the neck of a person, the windscreen of a lorry. You can look on this as plundering the sex organs of plants on a massive scale, plants that will never achieve the pollination for which they grew the gorgeous flowers: you can also consider the immense acreage that humans have given over to the cultivation of marigolds. There are many more marigold plants in the world than there would be if we humans didn't cultivate them, and the reason they are cultivated is because we humans find them full of beauty and meaning.

OPPOSITE AND ABOVE
Human delight: marigolds are grown, harvested and transported as an essential part of Mexico's great annual festival, the Day of the Dead.

OVERLEAF
Sustainability: good practice has preserved a rich ecosystem for humans and many other species in the Guassa grasslands of Ethiopia. Gelada baboons are among the beneficiaries.

The impact of human civilisation on the natural world was – or is – an event of devastating suddenness. Agriculture was invented around 12,000 years ago: and that's just a blink of an eye on the timescale at which evolution operates. The consequent alteration of the planet has taken place with breakneck speed. That speed has doubled and doubled again with the industrialisation of agriculture and of everything else. The Industrial Revolution has, in the course of a couple of centuries, changed our planet from top to bottom and made it unrecognisable. These changes have gone hand in hand with the galloping increase of the human population. It follows that all the plants and every one of all the other wild species are wonderfully well adapted – to the conditions that prevailed before humans took over. This, as it were, overnight alteration of the environment, this drastic shifting of the goalposts of evolution, has led to changes beyond computation all around the world.

Here's an example: the grasslands of the highlands of Ethiopia. This is a challenging environment for any form of life: it's classic four-seasons-in-one-day country. Here you will routinely get frosty nights followed by days when the temperature rises to the low 20s, and the thin atmosphere makes for bright and burning sunlight. The plants have evolved specialist strategies to cope with this: some have extra layers, some close up for the night and open again in the morning, some grow furry or waxy coats, some retain dead leaves as another layer of insulation.

The dominant plants are *Festuca* grasses; 60 per cent of the plants in the community belong to this genus, a group of tufted grass that are often

cultivated for gardens. It follows that if you are a herbivore, the *Festuca* grasses are the most obvious plant to eat. The plants counter this with sheer toughness: they are tough enough to survive these conditions and also tough to eat. Their stems are full of silica, which protects them from both the weather and the plant-eaters. *Festuca* may be the most abundant plants, but they are the last in line when it comes to palatability. As a further defensive ploy, they keep their nutrients and their growing points underground, inaccessible to most predators. Their pervasive root system binds the soil together and keeps it tied down, and this is essential for every other species that tries to make a living in this lofty place. The *Festuca* grasses can sustain predation from the gelada baboons and from the giant mole rats (sometimes called big-headed African mole rats). So here we have a natural balance: grasses whose toughness allows them to survive and thrive, so much so that they literally hold environment together.

Enter humanity.

The toughness of these grasses makes them second choice at best for most herbivores. But this toughness is just what makes them desirable for humans. The *Festuca* grasses make the best thatching; the grasses are also

ABOVE
A warden keeps watch over the Guassa alpine grasslands of Ethiopia: no one can harvest too much.

OPPOSITE
At home in the grass: a male gelada baboon in the Guassa grasslands of Ethiopia.

used for building, mattresses, robes, utensils, flooring, rope and as emergency grazing for domestic stock. All this makes these grasses highly useful, but in many parts of the highlands humans have got rid of it, using it unsustainably for thatching and other traditional purposes, or clearing it to grow crops or to make better grazing for the millions of cattle, sheep and goats. Without the network of *Festuca* roots, and with the constant trampling of domestic animals, the topsoil has been eroding at an increasing rate. As a result, poor soil is annually getting poorer. The ever-increasing human populations are faced with land that produces less and less food, and so people move into marginal areas and these too become devastated.

But not in the Guassa. Not in the Menz-Guassa Community Conservation Area. Here they have been looking after the grasses for 500 years, and what they have created – or rather what they have preserved – is an island of health and diversity in a sea of degradation. The people who visit this area of just 108 square kilometres – no one lives inside it – have a tradition of protecting the grasses. These days, rangers patrol the land with rifles. Poachers will come in to steal the grasses for their traditional uses, since there is more of it here than there is outside, or to dig out roots for fuel. But this

OVERLEAF
Room for many: an Ethiopian wolf, Bale
Mountains, Ethiopia.

practice is powerfully discouraged by the rangers, and the grasses thrive. Grass is the point of the protection but, since the grasses thrive, many other species of plants and animals are able to thrive as well, including the endangered Ethiopian wolf. The gelada baboons eat less of the grasses than they do elsewhere, because the system of protection allows other plant species to grow alongside the grasses, species more palatable to the herbivores. A human system that is designed to control human impact has allowed the Guassa to stay wild, and therefore to serve the needs of humanity.

It's all about human choice. Well, that's pretty well true of the entire planet these days. The invention of agriculture was the biggest single step in the history of humankind. It happened in many places in the world at more or less the same time and it changed forever the way we relate to plants. Before that, across the endless millennia, people had looked for plants. Now they started growing them. This was the moment that humanity took control.

ABOVE
Outdoor factory: irrigation circle for growing vegetables and ornamental flowers, Naivasha, Kenya.

OPPOSITE
Irrigation in progress: intensive potato farming in Norfolk, UK.

The principle is simple: instead of roaming about hunting for edible plants, you move the plants to a place convenient to yourself, one that combines shelter and fertile ground – in short, a farm. Humans stopped roaming and started to settle down. People had places of their own, fixed abodes, and, with them, opportunities to raid neighbours. It was the beginning of civilisation and the beginning of warfare; perhaps the two are inextricable.

The idea of agriculture is to grow the plants you like best, the ones that provide the best food and the most food. So you choose the seeds of the best plants when you're ready to grow your next crop. In a long and still-continuing process, one that takes place from one generation of plants to the next, these plants changed with the choices humans made. This is the process of domestication: you don't just tame them, you change them, and you do so by the process of selection. Charles Darwin called it artificial selection, as distinct from natural selection: the result of human choices, rather than the result of the pressures of the natural world. The plants that we grow in fields

are not much like their wild ancestors, the plants that humans first domesticated. A modern wheat variety is no more like the plant that was first cultivated in West Asia than a Yorkshire terrier is like a wolf. Some plants have taken very readily to life alongside humans, admirably suited to such co-existence. This is a process evolutionary scientists call exaptation: something used for a purpose other than the one it was evolved for. Humans hands evolved for grasping branches; they turned out to be very useful for using and making tools.

This is so much the case that many agricultural crops can't survive without humans. This point is illustrated vividly in the televised version of *The Green Planet* when David Attenborough shows us how sunflowers work. Time-lapse photography demonstrates how the great central disc of a wild sunflower curves as it ages, loosening the attachment of seeds to the plant and squeezing them out. The seeds tumble free with a gust of wind or at the alighting of a bird; they fall to the ground, and if all goes well, one or more of them will germinate the following spring, grow up as tall as its parent and produce seeds in its turn.

ABOVE
Stuck fast: David Attenborough demonstrates that cultivated sunflowers are unable to release their seeds. They can't reproduce without human assistance.

That is good news for the sunflower, but bad news for a human who wants to gather the seeds and consume their nutritious kernels. So Attenborough showed us the disc of a modern cultivated sunflower, the kind grown for cooking oil, cattle food and other purposes. It was as flat as if someone had used a spirit level, and it had retained all its seeds. These modern sunflowers are very tall, very fast growing, they produce enormous quantities of seeds – but they can't propagate themselves.

The most important staple foods of humanity are grasses: rice, maize, wheat, barley, and oats. Their wild ancestors also have seeds that drop off. Even more inconvenient for humans, sometimes the seeds then walk away and hide – and that's not an exaggeration. The wild oat produces spikelets that contain three or four seeds, each with two long bristles, called awns. The seeds fall to the ground in the natural way. The awns twist as they dry, and then untwist when they get wet. Tiny hairs on the awn grip the ground – with the result that the seed moves. It walks on two legs, wetting and drying, twisting and untwisting, and so it works its way along the ground until it meets a stone, which stops its progress, or a crack in the ground that it can drill into. Thus the falling seed finds a safe place, away from the predation of animals, and with luck, in the ideal spot to grow when the time comes.

ABOVE
Far beyond wildness: these cultivated oats have larger seeds
than their wild ancestors and lack the bristles that enable
wild seeds to bury themselves.

Seeds that fall off and walk away are no good for humans. So over the years, we have developed cereal crops with small awns and big seeds that don't drop off. This has taken place in a very short time in evolutionary terms, even if it seems a long time in the way that we humans understand history. It is nothing less than a revolution in the relationship between humans and plants – and by extension, between humans and the planet we live on.

You can look on it as a kind of deal. The plants agree to put all their energy into producing big fat seeds, wasting none of their resources in concealing them or dispersing them. Humans agree to look after everything the plants need: providing suitable ground, cleared and tilled and fertilised, and then planting the seeds in the best possible places. We have seen plants exploiting animals in previous chapters: the *Ceratocaryum* grasses fool dung beetles into burying them; many species hitch rides on the hairy heels of bison; there are endless examples in the great kingdom of plants. Here, plants exploit humans for almost all their needs, but by doing so they have sacrificed their independence.

The deal for humans is not remotely one-sided. Humans make possible the existence of certain plants; the plants do the same favour in return for

ABOVE
Taking control: growing bell peppers under glass in the Netherlands.

humans. A domesticated cereal plant has to grow and produce the right kind of seeds – but humans must do all the work that makes this possible. The end of the hunter-gatherer life was the beginning not just of agriculture but of work: real day-long back-breaking hard work. The great story of the Garden of Eden can be interpreted as a parable of the beginning of agriculture: humans leave the place (or time) of idleness and must now work every hour that God sends in order to survive, prosper and become ancestors. If you like, the plants have enslaved humanity, but that enslavement has allowed humans to dominate the planet as no species has ever done before.

Technological advances in agriculture have enabled us to produce more and more food, and so feed more and more people. It took about two million years for the human population to reach one billion; it took only 200 more to reach seven billion. In March 2020 the world's human population was 7.8 billion, expected to rise to 8.5 billion by 2030, 9.7 billion by 2050 and 10.9 billion by 2100. Around 40 per cent of all food produced is wasted, i.e. not consumed by humans or their domestic animals. About 80 per cent of all agricultural land is given up for livestock: so that we can consume most of our plants at second hand, through the medium of other species of animal.

The advances of agriculture have changed the nature of labour. In the developing world, up to 80 per cent of adults (and many children) work on the land; in the most developed countries, that number shrinks to less than two per cent. This drastic shift has come about because of immense farm machinery, and also by the use of synthetic herbicides, fungicides, insecticides and fertilisers; you don't have to hoe a field that's been drenched in glyphosates. Such farming methods are maintained by lavish subsidies from governments, and the way that subsidies are allocated directs the way that most agricultural land is managed. In 2005, the European Union began to phase out subsidies paid to farmers on the sole basis of the amount of produce; criteria based on wider environmental considerations have been brought in.

Further technologies continue to develop, most controversially with genetically modified plant varieties. In some countries, notably the United States, Brazil and Argentina, this trend is increasing, with potatoes, pumpkins, alfalfa, sugar beet, oilseed rape, maize, soy and cotton; in the European Union, GM crops are mostly banned. Other technologies have been used to increase food production: in many places, livestock, especially cattle and chickens, is kept permanently indoors, usually in crowded conditions; instead of grazing or foraging over many acres, the animals are fed in their barns on crops grown for the purpose. This is more economical of space: more cows on fewer acres. It is not the most efficient way of producing food; it is the most efficient way of producing meat. Some places have adopted vertical farming, that is farming on shelves, in which the production of food is not directly related to the acreage. There have been advances in soilless farming systems like hydroponics.

The tendency over the millennia, and increasingly over the last half-century or so, has been to create monocultures. Agriculture is about human control: it makes obvious, intuitive sense to get rid of every species apart from the one you are cultivating. You can find it in the vast expanses of prairie farming, and it is mirrored in suburban lawns, which are often treated with glyphosates to keep them free of pernicious things like daisies.

But there is increasing movement away from this line of thought. There is increasing demand for 'organic' food. That's an inaccurate term, since all growth is by nature organic; the phrase is used to mean a drastic reduction of invasive farming methods, especially the use of chemicals. The indiscriminate pursuit of monoculture affects, for example, the population of worms in the earth, creatures that aerate and process the soil and make it more effective. Monocultural farming reduces the number of worms in the soil, to the detriment of the growing medium and therefore, ultimately of the crops grown in it. In some countries, notably the UK, this is sometimes seen as a generational thing: 'young' farmers – in their 50s – are far less inclined to the idea that monoculture is the answer to everything.

Those with an interest in monoculture could do worse than visit Central Valley in California. Here, up to 80 per cent of the world's almonds are grown on 6,500 almond farms. Here, the dream of monoculture is as close as it has ever been to realisation. If it's not an almond, it has no right to be there – so it isn't. This is the perfect place to grow them: water falls in the form of snow on distant mountains, slowly released by the sun to flow into the valley during the long hot summer. Almond farmers have described Central Valley as a giant greenhouse: in the growing season it's always warm and sunny and the water only goes where and when the farmer chooses. In February, the trees come into flower: a million acres of the same blossom that inspired Vincent van Gogh to paint a masterpiece in France in the late nineteenth century.

A million acres of blossom – quite a thought. But there's a snag. Before an almond becomes an almond it must first be a flower. And in order to become an almond the flower needs to be pollinated. That's hardly unusual; every third mouthful we take requires animals to pollinate it. The snag is that down in Central Valley there are no animals to do the pollination. In order to achieve the monoculture, the place has been lavishly treated with herbicides and insecticides. The insecticides damage the beneficial as well as the harmful species. And just as importantly, if there are no flowering plants in the valley but almonds, what do the insects live on when their two-week flowering seasons is over? It's hard for any insect to make a living in this valley, and quite impossible for them to exist in numbers. There are nowhere near enough insects to cope with the annual fortnight-long emergency of pollination. That would be the case even if the decades of insecticide use could be put into reverse: the trees are planted so densely that the annual detonation of blossom would be too much for natural numbers of insects to pollinate them all. There are about 20,000 flowers on every tree, about 90 million trees adding up to 2.5 trillion flowers, none of which can produce an almond without an insect to carry the pollen from one flower to the next.

The answer lies in logical absurdity. The almonds are pollinated by domestic bees that are bussed in especially for the purpose. They come in on huge lorries from Colorado, Utah and elsewhere. You need two hives for every acre; each hive costs nearly US$200 to hire for the period of pollination. For these brief weeks the bees set about their age-old task: drinking nectar, taking pollen and bringing it back to the hive, and all the time inadvertently moving the male grains of pollen onto the female stigma in the flowers they visit, flowers that are gorgeous and conspicuous in order to be as attractive as possible to pollinating insects. After that, the hives are moved; other crops that use domestic bees for this form of pollination include alfalfa, clover (fodder crops), sunflowers, blueberries, watermelons and cucumbers. But the almonds dominate: this annual Central Valley operation is the largest controlled pollination event in the world. But almonds are merely the biggest example of controlled pollination: in an insect-depleted world, many other crops are in the same state.

There are increasing problems with this. There are suggestions that beekeepers are finding their hives up to one-third depleted after a visit to California. The reasons for this have not yet been established with certainty, but it's possible that it's to do with the chemical used to protect the trees, and it's hard to build the numbers back up. There are conflicting philosophies: to ramp up the intensity of the system with more bees, or to soften it by reintroducing native plants into the ecosystem, to support more wild pollinators and make the hives of domestic bees more effective. As it stands, Central Valley represents an extreme form of twenty-first-century farming, in which the natural forces that make growth achievable have been pushed to the limits of what is possible. The water is controlled through a complicated irrigation system, the pollination is controlled, and the plants themselves are controlled, and yet the system is close to breaking point. Is the future here, among ever-greater technology, intervention and increasingly invasive techniques? Or does it lie in a more relaxed and tolerant approach? Around 100,000 acres have been planted with self-fertilising almonds, a crop that can cope without insects. This is not necessarily an ethical question: it depends on how much farther we can push the system before it cracks, and also, on how much the human species will increase in numbers. These huge questions arise from a nut with nice depth of flavour that comes from the prettiest pink and white blossoms.

OPPOSITE
Farming on the limits of the possible.

TOP *Almond monoculture in California, USA.*

MIDDLE LEFT *Beehives must be bussed in; there aren't enough natural pollinators to do the job.*

MIDDLE RIGHT *A domestic honeybee visits an almond flower.*

BOTTOM *Almond trees stand to attention.*

We humans extend our love of monoculture everywhere we have an interest. We have never lost faith in our ability to take control of nature; as we have seen with the almond culture, when we find a problem we tend to adopt more extreme methods. And since we use wood for many things, it makes obvious good sense to seize control of the forests. Why let trees grow at random when we can take the decisions for them? We want the right kinds of trees in the places where it's most convenient to us to take them out and use them. We can produce a vast monoculture of wheat, in quantities that dominate a landscape – and we can do the same thing with trees: nice straight lines, all planted at the same time, all maturing at the same time, and all in the same place; exactly the species we need for our purposes, and no others to get in the way and spoil things. This makes good practical sense: sound economics. It's obvious: such practices are second nature to humans and we've got better and better at it. It's the way we have chosen to live; it's the way we have chosen to run the planet. What could possibly go wrong?

In Western Canada, the lodgepole pine produces just the right kind of wood for construction and for other purposes. It's a crucial plant for the economy of the area. You can see it over thousands of hectares: so regular and rhythmic that you can no longer really call it forestry. It looks like more like agriculture: certainly it is the cultivation of a crop rather than the nurturing of an environment. There are no inconvenient gaps, few trees of other species. These are trees as commodity: precisely what we humans want them to be; more like a Brobdingnagian wheatfield than a forest.

But there's a snag. The conditions we have created are exactly what the pine bark beetle has always dreamed of. It's as if we have managed a considerable part of the world for the express purpose of pleasing pine bark beetles. This is an insect that has built its lifestyle around a single species, that being the lodgepole pine. It's an eggs-in-one-basket strategy, but such strategies often work. If humans have a taste for the single species you have adapted for, you are in, as it were, clover, for as long as humans can make it work.

A female pine bark beetle moves in on a lodgepole pine. She will travel considerable distances to seek one out, but these days she doesn't need to. She bores into the bark. The tree has its defence ready: it produces resin that flows out and overwhelms the beetle. The beetle swallows what she can and passes it out through her abdomen. At the same time, she releases a pheromone that attracts other beetles. These arrive and also burrow into the bark. The tree continues to produce resin to deter the beetles, but some of them get through the defences and deposit their eggs. This relationship has continued across the centuries, the two species held in a dynamic balance: the trees mostly stay safe while the beetles mostly manage to produce the next generation.

The beetles that escape the clutches of the resin get beneath the bark, which is there for the tree's protection, and into the softer layer beneath. This is where you find the tree's circulatory system, carrying sugars down from the photosynthesising needles, and carrying water and nutrients up from the roots. In this layer, the beetles lay their eggs. As they do so they bring in a few spores of blue stain fungus. This fungus develops and starts to feed on the wood of the trees; by the time the egg has hatched into a larva, the tree's tissues are available to it as a nutritious gloop, brought into existence by the digestive processes of the fungus. It's one of the marvellous, almost incredible strategies that nature comes up with on a routine basis: the larva is locked in a cosy nursery with an endless supply of food. Thus the tree is attacked by a fungus that eats it, and at the same, the beetles are damaging its vascular system. It's a tough challenge, but over the centuries the trees have coped and the balance has been maintained. But things have changed.

The first problem is that there are now many more beetles. That's because there are many more lodgepole pines. When a tree is attacked, there are an awful lot of beetles to continue the process of damaging the tree's life-support system. But there is an additional factor, and that is climate change.

The larvae work on through the warmer months, devouring the bounty from the fungi, and towards the end of the summer they turn into pupae. They overwinter behind the bark in this form and emerge the following spring as adults. Many of these pupae are killed by the ferocity of those northern winters, but enough survive to continue the line. The problem is that with climate change more and more pupae are surviving and hatching out into adults, so there are more and more beetles ready to attack the trees the

PREVIOUS
*All is not well: a lodgepole pine plantation showing the damage
done by mountain pine beetles, Colorado, USA.*

following spring. And there is yet another problem: changes in the climate have given this area of Canada a relentless series of hot, dry summers. This depletes the resources of the trees and means they have less liquid – and therefore a reduced ability to make resin to protect themselves. Climate change means more beetles – and it also compromises the tree's ability to defend itself.

Climate change is a devastating business for every species on the planet, including the one that makes the decisions. But it is particularly severe on monocultures, for monocultures have no resilience. The lodgepole pine is a classic example of a planet with no Plan B.

The story of human civilisation is the quest for control: not to live in an environment but to create an environment. A great deal of the planet reflects the measure of our success: outside of Antarctica, around 77 per cent of the land surface of the planet has been converted to human use, mostly agriculture and tree plantations – and cities, of course. It is in our cities that you find the most extreme examples of human control – and, at the same time, the extent of our failure. Even in streets designed to hold only a single living species, a

ABOVE
Hard at work: larvae of mountain pine beetles in lodgepole pine, Wyoming, USA.

few non-human species find a way in, cocking a snook at human pride and human ambition, as it were.

This brings us to another classic David Attenborough *coup de television*, a great dramatic revelation of fecundity, one that brings to mind yet again the eternal Attenborough statement, one we have heard from the high Himalaya, a balloon in the stratosphere, a submarine in the benthic depths of the ocean or the world's most arid desert: 'Even here – there is life.' Perhaps this time the revelation was still more extraordinary, for it took place in Piccadilly Circus, the heart of London. The place is proverbial: British people will say of a hectic meeting, a crowded party, an unusual number of vehicles: 'It's like Piccadilly Circus.' Piccadilly Circus is full of humans. It is dedicated to their purposes and pleasures. Bright advertisements decorate the place and the sound of motors is never silent. And there was Attenborough at this most urban of places, this most citified part of a city, and he was showing us a wild plant. 'We call them, perhaps a little unkindly, weeds.'

Weeds. Plants that show us no respect, plants that grow where they choose, not where we choose, plants that mix themselves up in human life and refuse to die. They will do so even in cities, places where there is no soil, hardly any nutrients, very little water and what there is gets whooshed down the drain as fast as possible. And yet in these barren, human-filled places

ABOVE
A dandelion has germinated in a crack in a city street.

there are plants that get a toehold into the environment and live, produce seeds, and become ancestors.

Some plants are natural and almost inevitable colonisers of broken ground. Should fire, an earth slip, a fallen tree or some other calamity strike, such plants see an opportunity. They come in and grow fast, set seed fast and then die fulfilled. In the natural course of events, uninterrupted by humans, such plants will be followed by more solid and slower-growing species, in what's called the natural succession of vegetation, and it will proceed until it has reached its climax vegetation – which, in Piccadilly Circus, is a closed canopy oakwood.

It is unlikely that the vegetation of Piccadilly Circus will actually get that far – not for a few centuries, anyway – but in the meantime, these first arrivers, these pioneers, continue to sprout whenever and wherever they get the chance. The oldest cities in the world are lands for the pioneers, perpetually making a good start. And if you turn your back on them for a while, they will spread and start to create a micro-habitat.

So far as a plant is concerned, an old wall in a city is pretty much the same as a stretch of cliff. A plant that evolved for cliff-face life can make a go of life on a wall, given half a chance. A seed of ivy-leaved toadflax can germinate in a tiny crack or fissure and grow to flowering. A plant that is pollinated by day-flying insects must seek the light if it is to be found by its pollinators, but when this happy event has taken place and the flower has developed into a seedpod, the plant has an opposite challenge. It must now find the dark. And so the plant turns away from the light, seeking out the darkest places it can – like another crevice in the same cliff-face or wall. There it will plant its seed and, as the years pass, the wall, if unmolested, will become a toadflax garden, each plant planting its own seeds, generation after generation.

This is a good strategy for a plant in a city, because in such places, its stands to reason that most seeds that fall to the ground are doomed: crushed under feet and wheels, washed down the drain or simply landing on places where not even the most resourceful plant can make a go of things. But if you can produce a lot of seeds, you increase your chances. And if you can take gravity out of the equation, then you increase your range and with it, your range of opportunities. The sow thistle is a classic example: each seed is equipped with a downy parasol and, in the classic Mary Poppins manner, this acts not just as a parachute but as a flying machine. The seed can ride the

winds and rise with air currents and travel for miles. Floating seeds have been found a mile high up in the sky; they have been gathered hundreds of miles from their parent plants. They are classic pioneers, and if they can find a piece of broken ground, in a city or anywhere else, they rise again in their spiky glory. And, as any gardener will tell you, they are hard to get rid of.

A certain quality of resilience is a useful trait for any city-dwelling plant. There often seems to be a battle raging between humans and the opportunistic plants that surround us and our dwellings. A city is not the place to be a delicate flower. Survival in such antipathetic places is about toughness. The valerian has reputation of being almost indestructible. You can take a trowel to it and gouge it out with all the determination you can muster, but if you leave behind so much as a hint of a fragment of a root, the whole plant will regenerate: another of those plants that seems to mock human efforts at control.

ABOVE
Air miles: light micrograph of a sow thistle seedhead shows the parascending potential of each seed.

OPPOSITE
Putting down roots: a Chinese banyan tree finds a way to live in Hong Kong.

In the hot, moist tropics, rapid and eager growth is part of the nature of things – and that's true even in cities. Fig trees have evolved to grow in unlikely places. Some species send out a seed that germinates on another tree – any species will do – where it will then stretch out roots towards the ground and leaves towards the sun. Some species will establish themselves on cliffs and mountain walls, as we have seen with the living bridges of the Khasi people. But, in Hong Kong and many other tropical cities, figs will germinate on a building and miraculously make a living there. I lived in Hong Kong for four years and was accustomed to the daily incongruity of busy dwelling places festooned with washing and with a tree growing out of the wall. There is a helpful tradition of treating fig trees with a degree of respect.

And if we abandon any city building and let it fend for itself, it will become a garden in a remarkably short space of time. The pioneers will arrive as night follows day, and after that the bigger, long-lived plants move in. A sequence filmed for *The Green Planet* at an abandoned ironworks shows just how this works: a place that was built for the glorification of humankind is now a tribute to the glorious resilience of plants.

There is something admirable about the way some plant species won't take human domination lying down. Those that are opportunistic and fast-growing can often steal a march on humanity. But these traits can be dangerous. They can change ecosystems. They are doing so. We call such plants invasives, as if they came rampaging in like Attila the Hun. That hides the fact that they didn't really come in as marauders. They were often invited: humans liked them, took them from the places where they grow naturally and planted them elsewhere, often the other side of the world. The usual reason for this was that they look pretty. Other species came by accident, on ships and boots.

It's not a sure thing. Very few transported plants are able to make a go of things in an alien land. If you were to start a plantation of coconut palms in the highlands of Scotland, you'd be unlikely to cause an ecological disaster. There's a rough rule of ten per cent: only about ten per cent of alien plants will even survive in the wrong place, and only ten per cent of these will be successful enough to count as an invasive.

They do start with something going for them. There is a shortage of natural predators, natural competitors, parasites and diseases – that is to say, other species that have evolved to exploit them. They have a clean sheet in that respect. But the difficulties of an alien environment are immense. Such foreign plants find themselves in a habitat they did not evolve for. They have by definition a small population to begin with, so it's hard to find other plants of the same species for sexual reproduction. There are no natural pollinators or dispersers of seeds. Any invasive, no matter how successful it may become, starts off an underdog. Translocation can be understood as an extreme test of a plant's reproductive biology. Successful invasives across the world include kudzu, water hyacinth, Japanese knotweed, buddleia, *Desmodium* (as we saw in Chapter 1) and cheatgrass.

Also *Miconia*. This is not a familiar plant to most of us, but it's causing devastation in Hawaii. Locals speak of it as the green cancer or, for a change, the purple plague. In the years of increasing ecological awareness, Hawaii has acquired its own cliché: the extinction capital of the world. The place can be looked at as a laboratory model of human impact on the natural world: Hawaii was isolated for millennia, and was then subjected to devastatingly rapid development. The process began when the Polynesian people reached the place around AD 400, a brief enough time in terms of evolution. The pace was vastly accelerated when Europeans got to Hawaii in 1778.

The result of this combination of long isolation and hyper-rapid development is an extinction crisis. It's been calculated that 142 species of birds evolved in Hawaii and were found nowhere else: of these 95 are now extinct. The United States Fish and Wildlife Service lists 1,225 endangered species in the USA and, of these, 481 are from Hawaii; you could fit the landmass of Hawaii into Texas 25 times over. Hawaii has been described as the sounding board for the mainland; to be less parochial, we should also consider what such a crisis means for the entire world.

There are 1,400 species of vascular plants (so excluding liverworts, mosses etc.) on Hawaii; of these 90 per cent are found nowhere else. This extraordinary and unique diversity is threatened by the invasion of *Miconia*. Like so many other invaders, it was brought in because of its good looks: the plant is towering and imposing, with broad waxy leaves. It is native to Central and South America, where is grows rather sparsely. It was brought onto the island of Maui in the 1960s and 1970s and took a liking to the place. It has certain advantages: it grows fast and produces seeds in great quantities. These are wrapped up in tasty fruit, which many of the island's native (and invasive) animals seek out in preference to their traditional food. The plants are also able to fertilise themselves, so a single plant can make a good start on a place without need of assistance. The *Miconia* is tall by Hawaiian standards, where the forest canopy is seldom higher than 20 metres, so it is effortlessly dominant in the places where it has established itself. The leaves are large and shade out competition. The plant has disrupted the ecology of Maui, the second largest island of the Hawaiian archipelago. It has shallow roots that suck the water-table drier than it should be. It has come to an ancient and, in recent years, fragile habitat with the vigour of a bull in a china shop. *Miconia* has covered 80 square kilometres of Maui. Humans brought the plant in, and humans made its success possible by disturbing the environment and giving the invading plant a way in.

Humans made the mess; it is for humans to clear it up. But that's not a straightforward task. *Miconia* has taken over many steep slopes and cliffs striped with waterfalls in the dramatic scenery of this volcanic island. It is not easy country for humans to access, and even if it was, there are problems with contamination from boots on the ground, which can inadvertently bring in other unwanted items like seeds. The option of blanket spraying the *Miconia*

habitat with herbicide is not on: such indiscriminate spraying would cause more damage than the *Miconia* itself. But if indiscriminate spraying is no good, how about discriminating sprayers? And bringing them in by air?

This rather dashing project was invented and is run by James Leary of the University of Hawaii. The idea is to travel around the *Miconia* habitat by helicopter, armed with a paintball gun; you use the gun to fire a projectile filled with herbicide. You don't need much herbicide if you are meticulous and accurate: the active ingredient of these loaded paintballs is about the same as an aspirin tablet. Leary says that the technique is a 'high precision, high accuracy and surgical mentality'. You can travel about 30 metres from your target and place your paintball almost as carefully as if you were doing so by hand.

The project has been in operation since 2012 and has hit 15,000 targets in that time. It's slow going: perhaps three more decades will be required to complete the job. That's because comparatively modest funding only allows 120 hours of flying time in a year. Leary still finds the hunt-and-shoot part of

ABOVE
Weapon of conservation: Trae Menard, Director of Forest Conservation at the Nature Conservancy, fires a paintball gun to zap invasive plants with herbicide, Kuai, Hawaii.

it thrilling, but adds that the still greater pleasure is the return visit, when you can see the results: the native vegetation getting a chance again.

People are working increasingly hard to make sure that native vegetation gets a chance all over the world. That's because people are also damaging and decreasing the diversity of life of all kinds, in all kinds of different ways, and very few of us would get a not guilty verdict. That's a problem not just because it's a bad thing and a sad thing, but also because the strength and resilience of life on Earth depend on its diversity, as we have already seen many times in this book. And it's become resoundingly clear that we can't just sit back and hope for the best, hope that in due course there will be some kind of rebound. That dismaying truth is blindingly obvious in Hawaii, where extinctions are part of daily business. Many of the endemic plants on the archipelago are down to just a few individuals: without human intervention they will go extinct and their legacy of genetic diversity will be gone for good. In some parts of the state, the thing that has kept these plants going is their inaccessibility and their remoteness. That is good as far as it goes, but it makes it hard for humans to intervene in a positive way. Nevertheless, here and in many other places, dedicated seed-gatherers continue to perform marvels of courage and ingenuity to gather the seeds of endangered plants.

Seeds are extraordinary things. They hold everything a plant needs to know about life, the universe and everything, and yet a human can hold any amount of them in a single hand. Think of that: two dozen oak trees in your cupped hands, all in the form of acorns. Seeds come in a rich diversity that reflects the eternal inventiveness of nature: they take the shape of kidneys, squares, oblongs, triangles, eggs, discs and balls, some take to the air with wings, with parachutes, or as dust that floats on the wind, some float on water, some hook onto animals, some actively seek to be devoured by animals and some – like acorns – need to be buried by animals. Seeds in all their forms matter very much: they are the key to life.

So let us continue with a parable. A few years ago a Dutch researcher called Roelof van Gelder examined a leatherbound notebook. It was once the property of a Dutch merchant called Jan Teerlink, who visited the Cape of Good Hope in 1803. Let's spell that out: this was an object that was last in use more than two centuries ago. In its pages were seeds: seeds that had been cached there by Teerlink. They represented 32 species of plants. A few of

each of these were sent to the Millennium Seed Bank at Kew in London, and the people there tried to make them grow. Three of these species actually germinated. In other words, despite the two-century gap, during which the seeds were kept in far from optimal conditions, they still contained the spark of life. One of them, a plant called the tree pincushion, was reared to maturity; you can find plants of this species in the Temperate House at Kew Gardens, and when they bloom they are rather lovely things.

The moral of this is that if life can be preserved in a seed for two centuries in conditions that don't suit it at all – then it can surely hold on a great deal longer in absolutely ideal conditions. That is the thinking behind the Millennium Seed Bank: as we continue to do terrible damage to the diversity of life on the planet, so we can also keep at least some of it safe, as a genetic legacy cherished, cosseted and kept out of harm's way.

David Attenborough paid a visit to this bank, a bank that holds a greater fortune than all the normal banks of the world put together. He was able to marvel at the immense diversity of this impossible collection, looking at a specimen of *Furcraea parmentieri*. (The plant doesn't have a common name. The fact is that the diversity of plants is so immense that we haven't got casual, user-friendly vernacular names for them all, not by a long chalk. There are far more species than we humans can readily reckon. So we have to call this one *F. parmentieri*.) It's related to the agaves, which are succulents often grown in pots as houseplants. This species is found on the slopes of the Central Mountains in Mexico, where it can be damaged by wildfire. The plants are protected by the government, and they are now part of the great

ABOVE
Seeds of the future: left to right, horse chestnut or conker,
Cheiridopsis species from southern Africa, Norway maple
and woolly lousewort.

back-up plan at Kew. Should they go extinct in the wild, they will not be gone for good. What's more, what's a great deal more, is that *F. parmentieri* is not alone. It is one of thousands. There's life in the old seed bank yet, and will be for a good few centuries.

Seeds arrive there every day and from all over the world. They are sorted, cleaned, dried, X-rayed and sealed in airtight jars. From there, they enter the seed-vault, to be preserved for – well, who knows when they will be needed? It's a bit like the old style of fire alarm, the one that read, 'In emergency break glass'. Except that the emergency is already upon is. There have already been a number of cash withdrawals from the seed bank, and as a result species have been reintroduced into the wild. The contents of the bank are also used continually for research. Perhaps the best way of thinking of this is they are there and ready for when the world comes to its senses. It's a good job, then, that they are in it for the long-haul: a good job that two centuries can be considered as nothing more than a starting point.

The fact is that the Millennium Seed Bank drastically extends the possibilities of life on Earth for each of the species it contains, and perhaps for everything else that lives on the planet as well. So far, there are around two billion seeds representing 40,000 species from 190 countries. In times of despair, it is important to remember that no matter how great the extent of human folly, there are still people across the world working towards a wisdom that extends far beyond humanity. That's good because, in the end, our own survival depends on the survival of the world's plants. When a plant goes extinct unbanked, the knowledge it has accreted across the countless

millennia of its existence has been lost forever: lost to us, lost to the planet. The plants in the seed bank can be studied by scientists, their survival strategies, traits and adaptations informing our knowledge of life and the way we manage all plants, including those we use to feed ourselves.

The phenomenon of the Millennium Seed Bank excites two contradictory emotions. The first is an immense admiration for everyone involved in the project, along with deep gratitude that such a place should exist and be done so well. The second is a sort of baffled fury at the fact that we need such a place at all. How could it have come to this? How could we have let life on Earth reach such a point that we have to store bits of life in a vault because if we don't, they'll probably be lost forever? And perhaps more important: what can we do to make things better? No one at the Millennium Seed Bank would quibble at the idea that the world would be far, far better if we had no need for such a place. So how can we make this wonderful place redundant?

We all understand, to the point of horror, the idea of the extinction of animals. But the idea of plant extinction is not so easy to grasp. We are all deeply familiar with the idea of gardening and farming: you want a plant to

ABOVE
The world's bank: Millennium Seed Bank and poppy sculpture, Wakehurst, UK.

OPPOSITE
Priceless treasure: more than two billion seeds are stored in the vaults of the Millennium Seed Bank.

grow, you plant it and the plant gets on with the job of growing. Simple. So if a plant goes extinct in the wild, you just grow another one, perhaps taking a seed from the seed bank. You put the plant back where it belongs and bingo: you have put extinction into reverse. How wonderful it would be if it were all that easy. But gardening is not the answer to plant extinction, any more than a garden is the same thing as the wilderness.

What if the animals who pollinate the plant have gone extinct as well? What if the nearest population of potential pollinators is now five days' journey away? What if the animals who spread the seeds – by eating them and passing them out, by burying them, by carrying them on their fur – have gone for good? In the case of such an event, the plant either gets lucky and finds a substitute – or it doesn't. The world the plant had evolved for no longer exists, and in most cases extinction is the only possibility.

There are other, more subtle reasons that can make it impossible for an extinct plant to re-establish itself, no matter how benign the human support system. Conditions can change locally. There can be too much water or not enough, too much exposure to sunlight or not enough, changes in soil, changes in atmospheric conditions... A plant is a complicated thing, and the greater the complexity, the more things that can go wrong. It took a lot to stop

a typewriter: spill your coffee over your laptop and your day's work, and perhaps your life's work, is destroyed. And, as we have already observed more than once in these pages, the conditions in which plants grow on this planet have changed drastically in recent decades: the climate itself has changed and continues to change. Gardening is not the answer to plant extinction. The answer lies in the conservation and restoration of ecosystems. Or to put that in less scientific language, we need to cherish the wild.

And that is a truth that gives us an unexpected opportunity to rejoice. What's more, it's an opportunity to rejoice about rainforest. When did you last think of rainforest and cheer? When did you last feel happy instead of sad about the richest living environment on Earth?

Sebastião Salgado is a Brazilian and one of the world's great social-documentary photographers. He has repeatedly taken on the really tough assignments, the ones that make apparently impossible demands of your physical and mental resources. He covered the Rwandan genocide of 1994; in the course of 100 days, 800,000 people were slaughtered as the Hutu turned on the minority Tutsi. It was, as can be imagined, a devastating experience to live through. Even today, it is harrowing enough to look at Salgado's photographs from the warmth and comfort of home.

Home. Later that year, Salgado inherited the family farm. He returned home and found another kind of devastation. 'The land was as sick as I was,' he said. He needed, as we all do at certain points in our lives, to embark on the healing of harms. But the joyous place where he had grown up was now little better than a desert. There had been eight springs on the property that kept it lush and living; all of them had dried up. The river was polluted by a mining accident. This had been a childhood paradise for Salgado, a place where a boy could roam free and joyous. It had been 50 per cent forest; now it was 0.5 per cent forest. The trees had been cleared for cattle but now the place could scarcely support a single cow: it had been farmed into the ground. Many years earlier, the farm has been cut out of Atlantic rainforest. This, despite intense competition, is the most trashed habitat on Earth: 93 per cent of it has gone.

It was Salgado's wife Leila who started the process of healing. Let's restore the forest, she said. Let's bring back the old Atlantic rainforest to the 1,754 acres of this farm. It was an idea to bring a cynical laugh from the most optimistic person on the planet. The farm was dry. It was almost lifeless, apart from thick unmanageable swathes of invasive Africa grasses that had been planted to feed the cattle. It looked as burned out as Salgado felt. So they did it.

They gathered seeds from the broken patches of rainforest that remained. They established a nursery and began to rear trees and forest plants. They cleared the invasive grasses and began planting saplings. There were many setbacks. The project looked hopeless, ridiculous, the product of deranged minds. But they stuck at it, and the native plants began to establish themselves and the trees began growing towards the sky. Ten years after the replanting began, the springs came back. It was like a miracle, but it was nothing of the kind. It is the sort of thing that plants do.

The young trees brought rain back to the parched land. Trees release microbes and micro-particles that help the water in the atmosphere to condense and form clouds. Vegetation stores water inside itself, it collects water on its surfaces when it rains and it also enables the soil to hold more water that it otherwise would. When there are trees in an ecosystem, water goes into the atmosphere from evaporation from the soil and from leaf surfaces, and is also passed out from the plants by transpiration, a process we have already looked at. Trees make rain. I was reminded of the great children's book *The Promise* by Nicola Davies, which I read many times to my younger son: 'Green spread through the city like a song, breathing to the sky, drawing down the rain like a blessing.'

Which is the literal truth, nothing less. Rain came down on the Salgados' farm like a blessing, and the planting continued. The project was fine on its own, as an island, as a standalone statement, but it would be so much more significant if they could spread the word, spread the trees, spread the rain. The Salgados founded the Instituto Terra, to expand what they were doing into a teaching project. Local farmers, their attention caught by the return of water to the Salgados' farm, saw the point at once. They began to take on their idea of reforestation as an aspect of farming, as a way to make long-term and sustainable profits. So far 300 farmers have received support from the Instituto, learning how to plant trees that will protect the water supplies and safeguard the land.

The Salgados planted 2.5 million seedlings, representing 200 species of trees. They knew that the resilience of every ecosystem, but most especially rainforest, depends on its diversity. The erosion of the land by wind was halted as the young plants bound the soil. As the plants came back so did the animals that depend on forest. The pollinators returned the flowers to service; butterflies and hummingbirds lit the place up by day. The land animals came back: tamandua, coati, tortoise. Finally the predators returned as well: ocelot, jaguarundi, puma. The land has risen from the dead. Nothing less.

Here is a story of restoration, of the healing of harms. I have visited a similar project elsewhere in the former range of the Atlantic rainforest: the *Reserva Ecológica de Guapiaçu*. There I saw giant snipe and capybara in places where cattle once raised the dust; I rode a horse into the surrounding hills and looked down over a forest that now stretched to the horizon.

There is a moral here: despair is for wimps. It's not too late. Only very nearly too late.

So now, as humankind wobbles on the tightrope above the abyss of destruction, let us seek for balance. This is a hard thing to do when your subject is the state of the world. You have to tell the truth. You have to make the point that what's happening is serious, and you have to make it again and again. You have to make it clear that reckless and devastating destruction has become a central part of the way we run the planet, and that there are now so many of us humans that we are running out of space as well as time. But you must also point out the good things. That's as much a matter of honesty as it is of reassurance. The fact is that despair and false hope are both cheap options, the one as bad as the other.

It is hard to be an optimist these days, but nobody is asking you to be an optimist. And to be a pessimist is no way to live. Perhaps we need to set both of these ideas aside. Perhaps what matters is not whether you win or lose, but whether you are fighting on the right side. We need to look at the good projects, the hopeful projects, and we need to support them in any way we can, even if it's just cheering from a distance. To do so is not to blind yourself to facts and pretend that everything is all right: it is to open yourself up to the reality of hope.

So let us turn to Kenya. Like many places in sub-Saharan Africa, the local people depend on charcoal for fuel. This means that trees must be felled and burned. When that happens continuously the newly treeless land degrades and becomes useless for anything. As we have seen, fewer trees mean less rain. As human populations rise, so the demand for charcoal soars while the number of trees falls. It's a system heading for disaster. But an ingenious notion has been

OPPOSITE TOP
The Atlantic rainforest is the most trashed habitat on Earth. But here is the restored forest of the Instituto Terra in Brazil.

MIDDLE LEFT AND RIGHT *The gathering and planting of seeds at the Instituto Terra.*

BOTTOM *Even pumas have returned to the resurgent forest.*

developed to counter this, one that uses charcoal to bring trees back into the landscape. It is a most elegant contradiction.

It is about distributing seeds across the damaged land. Simply chucking seeds about would bring nothing more than short-term benefits to seed-eaters, both wild and domestic, but that's where the charcoal comes in. Balls of seeds are bound together with nutrients and then coated with charcoal dust. These can distributed across the degraded landscape and the animals won't touch them: the smell and the taste of charcoal is off-putting. The charcoal also allows the seeds to wait for the right conditions to germinate. Some of these will reach maturity.

The next step is to spread the seedballs over the landscape, and that challenge has been taken on with some eagerness. Anyone running an errand can plant a tree on the way. Wildlife rangers on poaching patrols can scatter them in the course of a day's work. They are thrown out by people in cars. Schoolchildren have set out on class walks to distribute seeds. More spectacularly, the speedballs have been dropped from paragliders and helicopters. More poetically, students from the University of Nairobi have developed a machine that fits on the back of camel; it spreads seedballs as the camel walks on.

In all, 13 million seedballs have been distributed since 2016. Of course this is long term project: the trees will need a couple of decades to establish a

ABOVE
Seeds of hope

LEFT *Rangers from the Masai Elephant Project disperse seedballs in the Masai Mara.*

MIDDLE *A double-handful of seedballs*

RIGHT *Seedballs starting to germinate.*

confident presence. The system is cheap and accessible to all and is being taken up all over the world. Right now trees are being planted in places where trees have been lost. Green is spreading through the degraded land like a song.

Life where it seemed life had ended, hope in places where hope had seemed impossible: it is impossible not to respond to such a spectacle. That is why *The Green Planet* television series ends not in in the wreckage of a lost rainforest or in the canyon streets of Shanghai, with its population of 24 million, but in a patch of ancient still living woodland in Britain. David Attenborough begins by telling us in uncompromising terms that the wood, 400 years in the making, is dying.

The Green Planet series is full of wonders. We have marvelled at the way plants adapt to the most extraordinary and apparently impossible challenges. But in recent years humans have changed the planet faster than forces of evolution can operate. These times of the warming Earth can be compared to another crisis of climate change that took place at the end of the Permian era 250 million years ago, this time caused mostly by volcanic activity. The Earth recovered all right: the snag is that it took 20 million years. The immediate future of the planet, then, depends on human choice. It to be hoped that we make the right choices as governments, corporations, large and small organisations and, essentially, as individuals.

This tract of ancient woodland that Attenborough visited is a small, lovely and vivid example of the right choice. The wood is taking damage along the

estuary that bounds it; about five trees are lost to erosion every year. The trees in the wood are being attacked from the inside by the invasive fungus Chalara that causes ash die-back; within 10 years, 30 per cent of the trees will be dead. Here, it seems, is a classic opportunity for despair.

But no. The local community has stepped in and begun to reverse the decline. There is now a pond, always a great way to summon life. A neighbouring field, once a place with less biodiversity than you can find in your kitchen, has been planted with 6,000 trees of 20 different species, Some have lived here before; others, like lime, may well thrive in a warmer climate. In every way, this is a project about the future. This is the wood that lived.

We take a deep delight in stories like this. It's one of the things that define our humanity. And we all know how to complete this book and the television series it accompanies in accordance with our great storytelling tradition: David Attenborough riding in on his white charger to defeat the bad guys and

ABOVE
David Attenborough in the New Forest. The choice we must make, he says, is 'to preserve and regenerate the greatest possible diversity of plant life on this – the Green Planet.'

make everything all right; David Attenborough to set about the healing of harms and bring us the planet that we deserve. Alas, as we all know, the truth is far more complicated than that, not least because we're all the bad guys and we've already got the planet we deserve. But we also know that every one of us can play the part of hero: we can all perform any number of small heroic acts in the course of our lives and become a force for good.

It was the Lone Ranger who rode to the rescue firing silver bullets. Sad to say, that's not an option for us. Instead, we need to face facts and act accordingly. We can do so by considering small things like an oak sapling just two years old, growing in that community woodland. It is a plant with an enormous possible future and equally enormous symbolic power. It won't be producing acorns for another 20 years, but that's no great matter, so long as humans give it the chance to grow. Given a good run, that little sprig of oak could live for 1,000 years, outliving us all, outliving our great-great grandchildren, growing on into a future we can only guess at. That's a tree and a thought to make us all pause: pause, and then accept the fact that there is no single solution. The solutions come in infinite numbers, they are local, right on your own doorstep, and they are within your scope. Every green leaf is important to all of us who live on Earth, no matter what species we belong to. We need to preserve the intact ecosystems of the planet; we need to make space for restoring ecosystems we have lost; we need to fill our working landscapes with a rich diversity of native plant-life – farmland, commercial forests, cities and gardens. We need to protect, we need to plant, and we need to start now. This book and the television series began with light: at the end the aim of both is enlightenment.

But as always, the last words must lie with David Attenborough. Filmed in Richmond Park, a short walk from his home and just a 20-minute train journey from central London, Attenborough summed up life on Earth once again in words to be savoured, treasured and remembered: 'Our relationship with plants has changed throughout history. And now it must change again. Whether it's what we eat, cultivate or whether it's what we like – we must now work with plants and make the world a little greener, a little wilder. If we do this, our future will be healthier and safer – and, in my experience at any rate, happier. Plants are, after all, our most ancient allies. Together we can make this an even greener planet.'

INDEX

ACKNOWLEDGEMENTS

Thanks to everyone at Penguin Random House, especially Laura Barwick, who organised the illustrations, the designer Bobby Birchall and copy editor Steve Tribe. I owe special thanks to Michael Bright, who coordinated the whole thing. Thanks as always to everyone at Georgina Capel Associates, especially to Irene Baldoni. And thanks for everything to Cindy, Joseph and Eddie.

Presenter and narrator
David Attenborough

Production team
Paul Barratt
Rupert Barrington
Lynn Barry
Miles Barton
Peter Bassett
Dawn Bruno
John Bryans
Matthew Clements
Adalean Coade
Isabelle Corr
Caroline Cox
Dave Cox
Beth Cullen
Jack Delf
Nathalie Dohrn
Lewis Easdown
Alison Fawcett
Lance Featherstone
Kate Fleming
Sandra Forbes
Donna Gomes
Jemal Guerrero
Michael Gunton
Katie Hall
Julian Hector
Danica Henderson
Georgia Hill
Melanie Hoffman
Tim Jeffree
Bridget Jeffrey
James P Lees
Kerrie Lloyd
Juliette Martineau
Ella Morgan
Jack Mortimer
Emma Napper
Maria Norman
Elisabeth Oakham
Emma Peace
Dan Prosser
Aicha-Louise Rakhdoune
Louis Rummer-Downing
Joseph Russell
Lisa Sibbald
Wilma Stefani
Rosie Thomas
Alistair Tones
Darren Tracey
Joe Tuck
Mitch Turnball
Joanna Verity
Natalie White
Nikki Wickham
Paul Williams

Camera team
James Aldred
Greg Allen
Shanu Babar
Tim Bailey
Adrian Barnett
James Bishop
Cameron Board
Roberta Bonaldo
Howard Bourne
Francis Brearley
John Brown
Keith Brust
Juan Pablo Bueno
Owen Carter
Song Chen
Patrick Cheung
Lloyd Douglas Clark
Mauricio Copetti
James Cox
Robin Cox
Tom Crowley
Trevor De Kock
Stefan Diller
Stuart Dunn
Jonny Durst
Robert Eder
Monika Ellinger
Neil Fairlie
Edson Faria Junior
Brett Foxwell
Russell Foulkes
James Frystak
Doug Gardiner
Justin Hadley
Micaela Hamilton
Neil Harvey
Robert Hollingworth
Kevin Hultine
Eric Huyton
Richard Ing
Stephen Kirkpatrick
Hugo Kitching
Gabriel Kocher
João Krajewski
Andrew Lawrence
Emilien Leonhardt
Sam Lewis
Graham Macfarlane
Lindsay McCrae
Rory McGuinness
Michael Male
Leonardo Mercon
Sam Meyrick
Hugh Miller
Oliver Mueller
Luke Nelson
Vikash Nowlakha
Kieran O'Donovan
Michael O'Shea
Nick Pitt
Joarley Rodrigues
Manu San Felix
Sealight Pictures
Matija Skalic
Sinclair Stammers
Justin Sullivan
Barny Trevelyan-Johnson
Ben Tutton

Darren West
Garath Whyte
Daphne Wong
Ignacio Romero Zurburchen

Director of timelapse photography
Tim Shepherd

Timelapse photography
Jacob Cross
Todd Kewley
Richard Mann
Jessica Mitchell

Visual Engineer
Chris Field

Engineers
Nigel Adams
Simon Knox

Scientific consultants
Robin Allaby
Anthony Ambrose
Alex Antonelli
Ian-Thomas Baldwin
Ulrika Bauer
Wendy Baxter
Jan Bechler
Barbara J. Bentz
Erica Berenguer
Peter Bruyns
Matyas Buzgo
James Carter
Martin Cheek
Colin Clubbe
Christopher Cook
Desert Laboratory
Mark Dimmit
Kingsley Dixon
Roy Fauntleroy
Andreas Fleischmann
Ryan Garrett
Simon Gilroy
Paul Green
Wang Hanbing
Jeff Kerby
Danny Kessler
Richard Kim
Bob Kipfer
Rupert Koopman
Richard Lansdown
Janet Lanyon
Donald H Les
Matthew Long
Thomas Lovejoy
Carlos Magdalena
John Manning
Cody Massing
Antonia Musso
Daniel J D Natusch
Ole Pedersen
Thomas Philbrick

Joseph Poore
Patrick Rogers
Rolf Rutishauser
Jasper Slingsby
Richard Unsworth
Orlando Vargas Ramirez
Alyssa Weinstein
Joseph White
Caroline Whitehouse
Ben Wilder
Tim Wilkins

BBC wishes to thank
Aja Akuna
David Alba
Carlos A. Lasso Alcalá
Maria de Carmo Amaral
Anza Borrego Desert State Park
APS GROUP
David Argument
Etwin Aslander
Clare Asquith
Steve Axford
Patrick Avery
Avon Wildlife Trust
Tim Bailey
Pedro Balonas
Craig Barber
Keila Pereira Barbosa
Adrian Barnett
Tarryn Barrowman
David Belton
Julia Belton
Abhra Bhattacharya
Volker Bittrich
Robbie Black
Black River Preserve, The Nature Conservancy
Steven Blackwell
Claudia Petean Bove
Nigel Bowers
Jo Bowman
Fabio Bozelli
Bristol Botanical Gardens
Christy Brigham
Oscar Castro Brizuela
Richard Brown
Jonny Brushett
Isabelle Bueno
Juan Pablo Bueno
Alim Buin
Steve Bunclark
Craig Burrows
Bushlife Safaris
Andrew Cairns
The Cactus Shop, Winkleigh
Cambridge Botanical Gardens

Natal Cândido
Cape Nature
Scott Carnahan
Allan Carrol
Carlos José Carvalho
Ivani Rocha Carvalho
Rick Chatham
Karen Cherrett
Steen Christensen
Cine Tirol Film Commission
Doug Clark
Alison C. Cohan
Freya Cohen
Hal Cohen
Martin Cohen
CONANP
Christopher Cook
Jon Cooper
Nigel Hewitt Cooper
Mauricio Copetti
Laura Coulson
The Council of the city of Stoke-on-Trent
Chris Crow
Ron Crump
Robin Cunningham
Adam Curtis
Ollie Deppert
Desert Botanical Garden, Phoenix
Linda Devlin
John Dickie
Charlotte D'Olier
Alejandra Douat
Roy Earle
Eden Project
Eleanor Edye
Rob Erasmus
Edson Faria Junior
Farm Studios
Robert Farthing
Fetcher & Baylis
Leigh Ann Ferreira
Kim Findlay
Fire Crews from the Greater Overberg FPA
Aaron Flesch
Forestry Commission (Forest of Dean)
Forestry England
Mr. Fourness
Brett Foxwell
Murray Fredericks
André Victor Lucci Freitas
Spencer Frost
Serena Fukushima
Mauro Galetti
Galleta Meadows Estate
Arturo Gamboa
Lisa Gammell

Reinard Geldenhuys
Christine Gemperle
Mark Gerasimenko
Getty
Luiz Gonzaga
Anoinette Goosen
Grootbos Nature Reserve
The Gudang & Yadhaykenu Traditional Owners of Cape York
Wondimagegn Hailemariam
Paul Hamilton
Chris Tan Hanafi & Oliver Deppert
Jon Hare
Olly Harrison
Craig Hastings
Hawaii Plant Extinction Prevention Program
Trica Oshant Hawkins
Patrick Hickey
Skip Hobbie
Oleksandr Holovachov
Jennifer Hunter
Instituto Terra
Intomedia
Tony Irons
Shernaz Italia
Olly Jelly
Janine Jones
Markwell Justin
Asrat Kahsay
Sirpa Kämäräinen
Sinita Kawasaki-Yee
Veit Kessen
Teddy Kinyanjui
Tadcha Klubcun
Nikki Knight
Rodney Knight
Jane Kokoch
Morningstar Kongthaw
Shiningstar Kongthaw
Jason Krueger
La Selva Tropical Research Station
David Lambie
Kevin Lanyon
Chien Lee
David Legros
Carlos Teixeira de Lima
Taylor Lockwood
Julio Madriz
Brooke Mahnken
Mandarin Film
Puja Maniar
Teódison Mano

Kelly Martin
Josephine Maxwell
Joe Mcauliffe
Helen McConnell
Kimberlie McCue
Lachlan Mcintyre
Stuart McTeare
Niki Meharg
Mendip Studio
  School
Getacher
  Mengesha
Kazimir Miculinić
Ivan Mikolji
Milford Haven
  Port Authority
Alen Milic
Alex Mills
Peter Mills
Mingo National
  Wildlife Refuge
Ministry of
  Information,
  Publicity and
  Broadcasting
  Services,
  Zimbabwe
Esta Mion
Alexander Mishin
Robert Moor
Martin Moore
Vanessa Rodrigues
  de Morais
Suzana Neves
  Moreira
Luisa Lima e
  Mota
Tohmondiam
  Mukhim
Marbel Muñoz
Christian Munoz-
  Donoso
Nick Murray
Helen Needham
Anna Niemela
Mai Nishiyama
Michael Oblinski
Ocotillo Wells
  State Vehicle
  Recreation Area
Tessa Oliver
Hank
  Oppenhiemer
Organisation for
  Tropical Studies
Osa Conservation
Ciara O'Sullivan
Verity Oswin
Yoichi Oyama
Pashley Manor
Instituto Homem
  Pantaneiro
Juan Pineda
Tyrone Ping
Mondew Pohtam
Pousada Rio Azul
  Lodge
Louis Price
Simon Pugh-Jones
Raul Puente
Angeloa Rabelo
Rainforest

Adventures
  Braulio Carrillo
Matt Ratson
Matthew Ray
Denis Rayen
Recanto Ecológico
  Rio da Prata
Sam Rees
Refugio Animal de
  Costa Rica
Owen Reiser
Richard van
  Rensberg
Gareth Reynolds
Rio Roosevelt
  Lodge
Roadrunner + Casa
  del Zorro
  Nursery
Roadrunner Club
Mark Roberts
Bruno Rocha
Flavia Rocha
Joarley Rodrigues
Jim Roth
Royal Botanic
  Gardens, Kew
Rajan Russell
Rolf Rutishauser
Charles Ryan
Sabah Park
  Department
Jose Sabino
Moris Munoz Salas
Lelia Salgado
Sebastião Salgado
Isabella Salton
Francisco Leiton
  Sancho
Lauri Sassali
Lorenzo Savergnini
Stefan Schnitzer
Sequoia and Kings
  Canyon National
  Park
Sequoia National
  Forest
Eunice Seravali
Kiku Severson
Shaw Nature
  Reserve
Andy Chia Chee
  Shiong
Nicole and Don
  Shearer
Tamara Sherrill
Tetsuo Shimada
Shutterstock
The Sloth Institute
Robert Smith
Danielo José Lima
  de Souza
Kira Steiner
Andrew
  Stephenson
Josh Stratton
Studio Salgado
Mirjana Sučić
Maciej Sudra
Anna-Riitta
  Sulander
Chris Tan

Tim Tchida
Terra Folia
  Biosphere
  Reserve
Janet Terry
Thaimaçu Lodge
Patrick Thrash
Ian Thwaites
Geovani Tonolli
Kevin Townsend
Triplefin Media
Reetta Turula
Teija Turunen
Seija Väisänen
Danielle Veenstra
Robert A Villa
Aidan Vincent
Katja Vira
Marcus Vira
Waikamoi
  Preserve, The
  Nature
  Conservancy
Shan-Shan Wang
Eva Wegel
Kebwezi Well-
  wishers
Haylie Wells
Wells Gray
  Provincial Park
Louise Wessels
Andrew White
Andrew Whitworth
Reg Williams
Kate Wilson
Stephen Woods
Nicholas Wray
Carlos Adriano
  Ximenez
Steve Yanoviak
Yayasan Sabah
  (Sabah
  Foundation)
Wang Yishan
Sylwia Zbijewska
Christian Ziegler

Editors
Steve Barnes
Will Brown
Peter Brownlee
Nick Carline
Jessica Hallier
Caroline Hamilton
Emma Jones
Tim Lasseter
Robin Lewis
Dave Pearce
Owen Porter
Jack Roberts
Sam Rogers
Tom Rowe
James Taggart
Steve White

Original music by
Benji Merrison
Will Slater

Music performed by
Hungarian Studio
  Orchestra

Post production
Dave Cawte
Miles Hall

Online Editor
Wes Hibberd

Colourist
Simon Bland

Sound Editors
Kate Hopkins
Tim Owens

Sound Recordists
Sean Millar
Gary Moore
Chris Watson
Andrew Yarme

Dubbing Mixer
Graham Wild

Visual Effects
Moonraker

**Additional
graphics**
Henrique Bispo
Christopher Jones

**Commissioning
Editor for BBC**
Jack Bootle

**BBC Studios Sales
and Distribution**
Patricia Fearnley
Monica Hayes
Esther Lawrence
Mark Reynolds
Elena Sollazzo
Amy Stone

**Academic
consultants for The
Open University**
Yoseph Araya
Julia Cooke
David Gowing

**Co-production
partners**
bilibili
France Télévisions
PBS
ZDF

10 9 8 7 6 5 4 3 2 1

Witness Books, an imprint of
Ebury Publishing
20 Vauxhall Bridge Road,
London SW1V 2SA

Witness Books is part of the Penguin
Random House group of companies
whose addresses can be found at global.
penguinrandomhouse.com

Penguin
Random House
UK

This book is published to accompany the
television series entitled *The Green Planet*,
first broadcast on BBC One in 2022.

Executive producer: Mike Gunton
Series producer: Rupert Barrington

First published by Witness Books in 2022
www.penguin.co.uk

A CIP catalogue record for this book is
available from the British Library

978-1-785-94553-3

Commissioning Editor: Albert DePetrillo
Project Editor: Michael Bright
Picture Research: Laura Barwick
Image Grading: Stephen Johnson,
www.copyrightimage.co.uk
Design: Bobby Birchall, Bobby&Co

Printed and bound in Germany by
Firmengruppe Appl

Penguin Random House is committed to
a sustainable future for our business, our
readers and our planet. This book is
made from Forest Stewardship
Council® certified paper.

PICTURE CREDITS